DIRT RICH

THE GROWER'S NO-TILL HANDBOOK FOR ABUNDANT SOIL,
PROFITABLE CROPS, AND A REGENERATIVE FUTURE

HARPER MOSS

Feel free to join our mailing list for updates and information about other upcoming books by

HARPER MOSS

Table of Contents

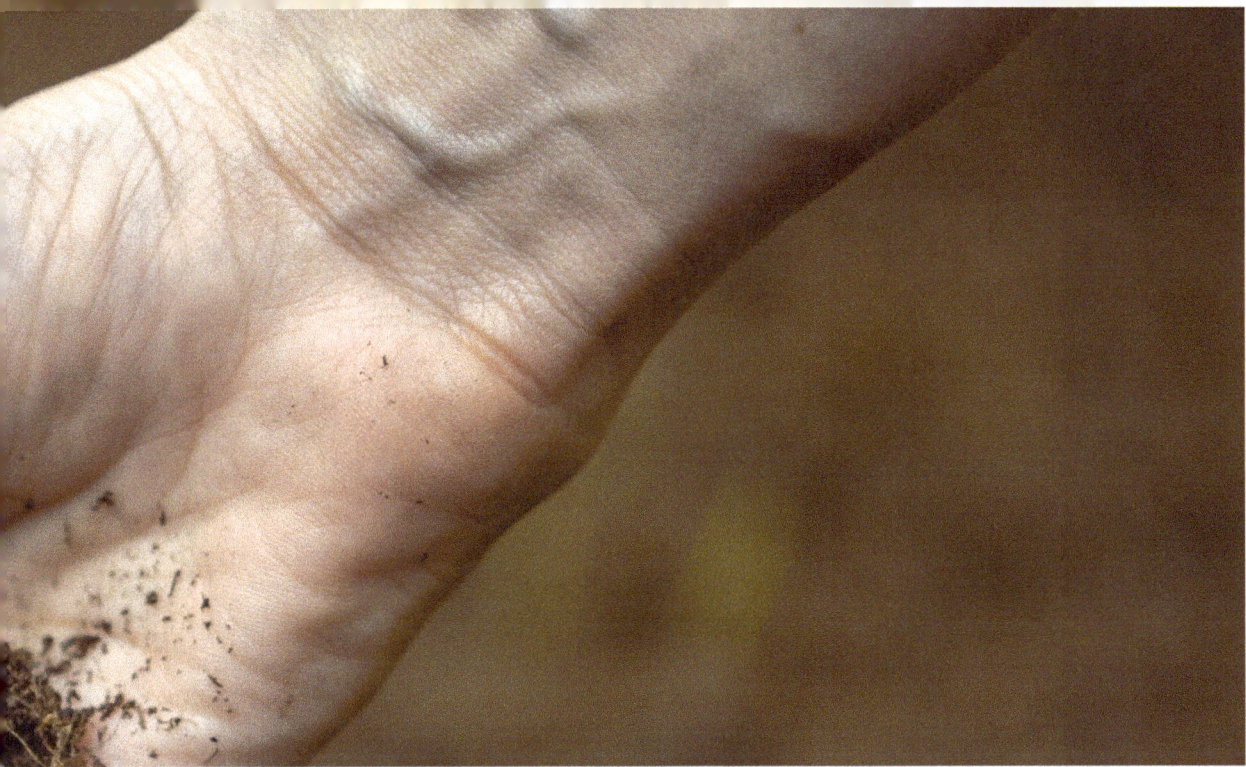

Introduction: Soil is Not Dirt— It's the Beginning of Everything

A Typical Journey

The majority of farmers do not begin with a soil-first mentality. There is frequently a sense of urgency at the start of the journey: produce food, turn a profit, keep the weeds out, and repeat the cycle. Decisions are made based on this urgency: herbicides to protect the crop, fertilisers to increase yields, and tilling to "fluff" the soil. Conventional farming implements turn into routines rather than enquiries. Instead of a guidebook for observation, you are given a recipe.

Growers I have met get caught in this cycle for years. Their yields are irregular, their inputs are more expensive, and their soil gets tougher. The land used to produce with vitality, nevertheless. There is a missing element. In actuality, that something is soil health—not just in name. Slowly, perhaps out of curiosity or annoyance, the realisation dawns. After avoiding tillage, a grower observes an increase in earthworms. It appears that a cover crop completely alters the atmosphere of the field. Water intrusion gets better. Crop resilience increases. Instead of acting as a passive substrate, the soil starts to act like a living entity. The voyage changes at that point.

What Makes This Book Different

The majority of agricultural guides discuss profit margins, equipment, and yield. The soil is the primary focus of this book, and not just as a resource, but also as a partner. There are no quick-fix formulas available. You will discover long-term thinking that is grounded on biology, observation, and useful tactics that capture the intricacy of actual farms. No-till is a technique to understand systems, not just surfaces; it is not a panacea.

Not only are activities being described here, but your perspective is being reframed as well. Composting, mulching, cover crops, and rotations will all be discussed, but always from the perspective of how they benefit the living soil. You will discover how to recognise both health and decline indicators in your crops. Due to the fact that regeneration is not about perfection, you will hear tales of both success and failure, including mine. Engagement is key.

Who This Book is For

This book is for growers who wish to improve their soil and their financial situation, regardless of how many acres they manage—one or 1,000. Perhaps you are a market gardener trying to save money on inputs. Perhaps you are fed up with compaction and erosion as a row-crop farmer. Perhaps you are an instructor guiding the next generation of land stewards or a homesteader cultivating for your family.

A background in soil science is not required. Curiosity, an openness to observation, and a desire to go beyond extraction and towards cooperation with the land are necessary. This book implies that you are tired of working against nature rather than with it, rather than that you are terrified of work.

What You'll Learn

You will learn about the microbial networks, fungal highways, and carbon cycles that make up the life under your feet. By employing techniques that preserve moisture, repair structure, and lessen erosion, you will discover how to create soil rather than only use it.

You can anticipate thorough explanations of plant-soil interactions, compost construction, nutrient cycling, and no-till techniques. You can anticipate field notes, experiments, case studies, and sidebars featuring perspectives from many agricultural fields. Above all, anticipate a method that honours complexity without becoming incomprehensible. You will be prepared to act in a way that suits your situation, not someone else's.

How to Use This Book

Depending on your needs, you can either read this book from beginning to end or skip around. Though it focusses on different topics, each chapter builds on the one before it. Use field notes and sidebars, which offer glimpses into actual farms, to compare, challenge, and investigate concepts. To assist you in putting changes into practice, the drawings and soil health diagrams provide visual signals. Each chapter concludes with practical steps you may take immediately and additional reading recommendations for in-depth discussions.

Additionally, this book is meant to develop alongside you. Every season, go back to it. After a year of composting, go over the soil biology chapter again. With the cover crop section, reevaluate your rotation plan. Allow it to join you in developing a relationship with the land.

Callout: "What Makes Dirt Rich Unique"

This book does not promote no-till as a fad. It investigates no-till as an ecologically based ideology. Every industry is unique, and "Dirt Rich" acknowledges that the finest techniques come from trial and error, observation, and feedback. Here, data-supported techniques are used, but there is also opportunity for subtlety. This is an ongoing conversation with the earth, not a set of directives.

This book is unique because it combines practical application with serious soil science. It blends the most recent findings with the firsthand accounts of growers who are pursuing this route. It appreciates failure tales just as much as innovations. The objective is a manual for managing soil as the basis of life and livelihood, not only a guide for cultivating crops.

Field Note: My First Mistake with Tilling

I used to think tilling was progress. My first growing season, I rented a heavy walk-behind tiller, eager to make my plot "ready." The ground was compacted and rough, and the tiller fought me the whole way. I remember the smell of raw earth, the churned-up weeds, the feeling of accomplishment—brief as it was.

Within two weeks, the soil crusted. Weeds surged back faster. My seedlings struggled. The ground dried out even after a good rain. I had broken the soil's back and called it success.

That first mistake taught me a painful lesson: just because you can till doesn't mean you should. I had disrupted soil structure, wiped out fungal networks, and invited erosion. I wasn't preparing soil—I was destroying habitat. That failure was the beginning of my education.

Over the years, I traded that tiller for broadforks, cover crops, and mulch. I started measuring soil respiration, not just surface texture. I learned to observe—truly observe—the life beneath the surface. That shift didn't happen in a season, but it began with one mistake that revealed how little I understood. Mistakes are often the best teachers when you're willing to listen.

Soil is not dirt. It's alive. It's intelligent. It remembers what you do to it, and it responds. This introduction is your invitation to reconsider what it means to grow food, to care for land, and to shape a future where farming restores instead of depletes. The rest of this book will give you the tools, insights, and stories to deepen that journey. Welcome to a new way of seeing the soil—not as the end of a shovel, but as the beginning of everything.

PART 1:
The Roots of the Revolution — Foundations of Living Soil

Chapter 1:
What Is Living Soil, Really?

The Soil Food Web Explained

Living soil is an ecosystem brimming with life, not only minerals and organic stuff. There are billions of species in every handful of good soil, ranging from earthworms and arthropods to bacteria and fungi. Scientists refer to the dynamic, interconnected system in which these organisms interact as the soil food web. Each participant affects soil fertility, structure, and the general health of the land.

Microbes, including bacteria, actinomycetes, and fungus, are at the base of the web. By breaking down organic matter, these tiny powerhouses release nutrients in forms that plants can absorb. Mycorrhizae, symbiotic interactions between fungi and plant roots, aid in the uptake of trace minerals, water, and phosphorus by plants.

Protozoa and nematodes are found above the microbial layer. These microscopic organisms release nitrogen and other nutrients in forms that plants may use by feeding on bacteria and fungi. The arthropods, which include beetles, mites, and springtails, each contribute to the destruction of organic matter and the promotion of microbial activity. Perhaps the most well-known soil inhabitants are earthworms, which mix organic matter into the deeper soil layers and improve drainage and aeration by tunneling into the soil.

A constant cycle of decomposition and regeneration is guaranteed by this web of life. In addition to growing in soil, plants also interact with it by supplying bacteria with carbohydrates through root exudates. Microbes provide nutrition, prevent illness, and act as a buffer against environmental stress in exchange. Understanding the soil food network makes you want to assist nature rather than try to dominate it.

Soil as a Superorganism

Healthy soil behaves like a living thing rather than a machine. It eats, digests, breathes, and recovers. It sustains a variety of lifeforms, responds to stress, and heals from wounds. Because of this, a lot of growers and soil scientists now call soil a superorganism, a self-organizing, self-regulating system composed of both living and non-living elements.

The skeleton of soil is its physical composition, which includes the proportions of sand, silt, clay, and organic materials. The microbial life functions as the organs, performing vital functions such as immunological protection and nutrient cycle. All components of the ecosystem receive resources from the chemical components that make up the blood.

Tilling soil damages its "skin" and exposes its "guts" to the weather. Overfertilization suppresses its internal processes and throws off its hormonal balance. Soil loses its protective layer when we leave it bare, which leads to carbon bleeding.

Everything changes when soil is viewed as a superorganism. It is now a collaborator rather than a passive growing medium. It flourishes and gives back if you take good care of it. It disintegrates if you handle it like an inanimate thing. Not only does the soil sustain life, it is life itself.

Myths About Dirt vs. Soil

What is under your fingernails is dirt. The world is fed by soil. That difference is important. However, we use the terms interchangeably in ordinary speech. Semantics is only one aspect of the issue; another is a cultural disconnection from the land.

The dirt has died. It is what is left over after soil loses its organic substance, biology, and structure. Water is not well absorbed by dirt. It is easy to condense. It does not produce much without a significant amount of outside input, and it does not smell particularly good.

In comparison, soil is living. It has a varied population of organisms, structure, porosity, and moisture retention. With help, it is capable of self-regeneration. People are not being lyrical when they claim that their farm is merely "a bunch of dirt." They are indicating that the system has been exhausted.

It is like supposing a corpse is simply a sleeping person if you assume soil is just earth with fertilizer applied. The difference is life. And the ability of soil to sustain crops, hold onto water, and fend off illness is due to that life—microbial, fungal, and invertebrate.

Tilling, naked fallowing, and synthetic-heavy management cease to make sense once you make that mental change. Instead of trying to control the soil, you try to create it.

Myth vs. Reality: "You Can Fix Any Soil with Enough Fertilizer"

This is a common—and harmful—myth. The notion that bad soil can be "fixed" by applying fertilizer disregards all of the knowledge we have about soil biology. While synthetic nitrogen and phosphorus fertilizers, in particular, can momentarily increase yields, they have little effect on restoring the health of the soil. They frequently intentionally damage it.

Excessive fertilizer use causes nutrient lockout, acidification, and salt accumulation. It increases reliance on outside inputs, promotes shallow roots, and decreases microbial diversity. Yields eventually decrease unless additional fertilizer is supplied. A cycle of diminishing rewards is at play.

Restoring the life of the soil through the addition of organic matter, the encouragement of beneficial bacteria, and the reduction of disturbance is what actually improves bad soil. Cover crops, mulch, compost, and no-till techniques all aid in restoring soil equilibrium. Often, the soil already contains the nutrients you require; they are simply locked up. The secret to unlocking them is biology.

It is like feeding a corpse when you add additional fertilizer to dead soil. You are not resolving the fundamental problem. As the system continues to deteriorate, you are concealing symptoms.

Fertility is biological in nature. The nutrients will come if you create the life. The plants will take care of themselves if you develop the biology.

Field Note: My First Time Seeing Compost Steam

The first time I moved a compost pile and watched steam emerge from the center will always stick in my memory. I had never made compost before, it was early spring, and the air was still chilly. I had combined some manure, straw, garden trimmings, and food scraps. I had low expectations.

However, it was warm when I inserted the fork into the mound that morning. Hot, not lukewarm. A plume of steam curled into the cold air as I raised the forkful. It smelled sweet and earthy, like rain in a forest. I realized then that this pile was living.

It was more than just decaying. It was changing. The bacteria inside were forming humus, creating structure, producing heat, and breaking down complicated molecules. That compost mound was a living force for regrowth, not waste.

My perspective on organic matter was altered by that encounter. I began to view "waste" as nourishment for the earth rather than something to be thrown away. It was evidence that life breeds more life because of the heat, steam, and odor.

I gave up attempting to make my fields better from the outside in that day. I began to nourish them from within.

There is no metaphor for living soil. For regenerative gardeners, it is a daily experience and a scientific fact. The first recycling system found in nature is the soil food web. The superorganism paradigm serves as a reminder that soil is not a chemical formula but rather a biological totality. By busting misconceptions about fertilizer and soil, we can adopt methods that create resilience from the ground up.

The foundation for all that comes after is laid forth in this chapter. If nothing else, keep in mind that cultivating, protecting, and honoring healthy soil is more important than creating it. It is not your job to control it. Making it come to life is your responsibility.

Chapter 2:
Why No-Till Works
When Tillage Doesn't

Why Tilling Fails (and Keeps Failing)

Agriculture has used tillage for thousands of years. It has always been regarded as an indication of production, from the earliest oxen-drawn plows to the enormous machines of today. In addition to smoothing the area and burying weeds, it also stirs the soil and facilitates planting. The dirt appears tidy, bare, and "ready." Many growers find solace in that image.

However, there is long-term devastation behind that seeming order. The very mechanism that enables soil to support life is upset by tilling. Tillage breaks up soil aggregates, exposes organic materials to fast oxidation, and eliminates fungal networks that develop over many years with each pass. It pulverizes the soil surface's protection, making it susceptible to erosion from rain and wind.

Furthermore, it does not end there. Water infiltration is decreased by tillage. It disrupts the natural sponge-like texture of the soil. Till farmers frequently observe runoff, inconsistent wetness, and puddling in their crops. More tillage is necessary when weeds reappear stronger. Below the tilled layer, compaction forms a hardpan. Despite their initial appearance, yields will eventually start to decrease unless inputs are increased.

Tillage causes issues that seem to necessitate further tillage, which is the circle that binds so many growers. It is a treadmill. Even though you are running faster, you are not making any progress.

No-till farming is more than just not tilling; it is a deliberate choice to break the damaging cycle and restore what tillage destroys.

Soil Structure, Carbon Loss & Biological Disruption

Structure is a feature of good soil. It is rich in organic materials, crumbly, and airy. Its aggregates, which are collections of organic matter and mineral particles bonded together, are essential. They provide water and air room to flow. They shield microorganisms and roots. When soil life is flourishing, they naturally form.

These aggregates are destroyed by tillage. When they break apart, the organic stuff within is exposed to the air, which causes carbon dioxide to be released and microbial consumption to occur quickly. That is quantifiable, not theoretical. Carbon is lost from tilled crops more quickly than it is gained. They get compressed, brittle, and depleted with time.

This carbon loss is a disaster in agriculture as well as an environmental one. Soil fertility and structure are attributed to carbon. Microbes use it as a food source. It is the nutrition cycle's engine. The ability to produce resilient crops without the need for external inputs is lost when it is lost.

Disruption of biology is just as harmful. Tillage destroys mycorrhizal fungi, which create extensive underground networks that link plant roots. Particularly in soils with poor fertility, these fungi are essential for the uptake of nutrients and water. When they go, plants have to put in more effort and rely more on fertilizer.

Additionally, tillage kills insect habitats and harms worm populations. It promotes weed development and quick nutrient loss by changing the soil food web from a stable, fungal-dominant system to a bacterial-dominant one.

Tillage essentially upends a system that is physiologically balanced.

No-Till vs. Low-Till: Which is Right for You?

Not all farms are prepared to switch to a no-till system overnight. It is alright. Progress, not perfection, is the aim. With good planning, no-till means preserving ground cover, causing the least amount of soil disturbance, and promoting soil biology. It is most effective when used in conjunction with mulching, composting, and cover crops.

In contrast, low-till permits sporadic disturbance. To control weeds or break up crusted soil without causing significant disturbance, some producers employ techniques like shallow surface cultivation or rotary harrows. Low-till may be used as a compromise or transitional measure before implementing complete no-till in some crop systems or regions.

Intention is crucial. Are you causing the least amount of soil disturbance possible? Are you creating more organic matter than you are destroying? Are you using biology instead of force to manage your system?

No-till is excellent in maintaining soil moisture, microbial life, and structure. Perennial systems, vegetable operations with permanent beds, and row crops bolstered by cover crops and residue management are frequently better suited for it.

In wet springs where the soil remains too chilly and compacted, or on land that has been degraded by years of intensive tillage, low-till may be required. A broadfork, shallow ripper, or focused disturbance may be helpful in those situations—but only if the intention is to restore soil life.

When the mentality changes from dominance to cooperation, both systems function.

Common Grower Question: "Can I Use a Broadfork and Still Be No-Till?"

Yes, you can still be no-till even if you use a broadfork. It is actually one of the greatest resources for many producers looking to switch to a no-till system. The dirt is loosened with a broadfork without layer flipping or inversion. It preserves the natural horizon lines of microbial and fungal life, known as the biological strata.

The purpose of inserting a broadfork and gently rocking it back is to aerate the soil rather than disrupt its structure. The biology is mostly unaltered, roots can grow deeper, and water infiltration is improved. This type of disturbance is respectful and works with the architecture of the soil rather than against it.

A broadfork is a common tool used by growers to prepare beds early on, particularly when transferring a grass or compacted field. The structure is then maintained open and biologically active through the use of compost, mulch, and cover crops. The broadfork subsequently stops being a seasonal crutch and instead becomes an infrequent instrument.

How and why you use it is what counts. You are doing no-till both in theory and in practice if your aim is aeration rather than cultivation and your beds remain undisturbed for the majority of the year.

Here, it is crucial to stay away from orthodoxy. Purpose is what no-till is all about, not purity. A tool has a place in your system if it aids in improving soil quality without destroying microbiological life.

Although tillage may have played a role in farming's history, it need not be a part of its future. It fails because it is harmful over time, not because it is ineffectual in the near term. It disrupts the processes that create fertility, putting growers in a vicious cycle of erosion, compaction, and chemical reliance.

No-till provides an alternative. It produces organic matter, safeguards biology, maintains structure, and promotes long-term resilience. The important thing is to stop fighting the soil and start listening to it, regardless of whether you start with low-till or go completely no-till.

Every plow pass is remembered by the earth. However, it also retains memories of every inch of mulch, every cover crop, and every moment it is allowed to grow unhindered. Decide the memory you wish to preserve.

Chapter 3:
Know Thy Soil — And Work With It

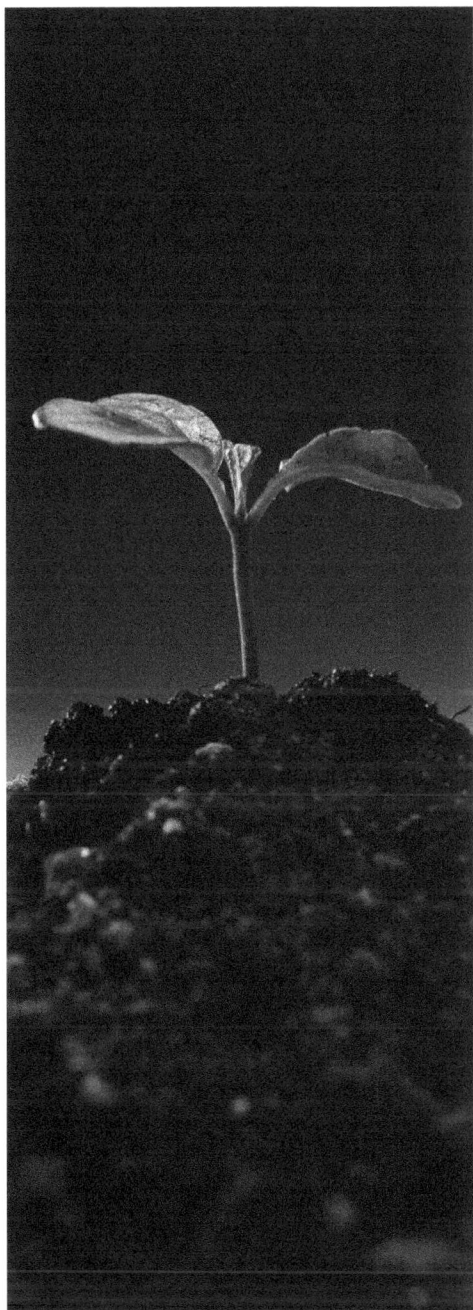

How to Test (Lab, DIY, and Digital)

Understanding your soil entails letting go of presumptions. It implies the collaboration of facts, observation, and experience. By testing your soil, you can learn more about what is going on underneath your crops, including what nutrients are there, what isn't, how much organic matter is present, and how biological life is thriving.

The most thorough option is lab testing. Samples are collected, sent to a soil lab, and data are obtained on pH, cation exchange capacity (CEC), organic matter percentage, nutrient availability (NPK and micros), and occasionally even microbial activity upon request. When setting up a baseline, doing troubleshooting, or switching to regenerative methods, these tests are crucial. Select a lab that specializes in mineral balance or soil health measurements that go beyond traditional fertility panels; some specialize in regenerative systems.

Do-it-yourself testing techniques are also beneficial. You may get overall pH and main nutrient readings with easy-to-use at-home kits. Physical soil tests, such as the earthworm count, slake test, or infiltration test, are even more useful.

These techniques let you know how well your soil drains, breathes, and sustains life. They teach you to think biologically, are inexpensive, and do not require any particular equipment.

Soil analysis is becoming more accessible thanks to digital instruments. Instant readings can be obtained via web-based platforms, smartphone-connected probes, and portable scanners. These tools assist you in tracking changes in real time, but they do not take the place of labs. Applications have the ability to document field observations, compare your data with satellite imaging, and recommend amendments or cover crops. Combining high-tech equipment with practical experience is crucial.

Testing is used to gauge progress as well as to detect issues. The changes in healthy soil are gradual. Testing keeps you focused and patient.

Reading Soil with Sight, Smell, and Touch

Hands and noses existed before lab findings and applications. One of a grower's most effective tools is still observation. There is no need for equipment; sight, smell, and touch can all provide information about the health, condition, and any issues of the soil.

Let us start with color. Black or dark, rich brown soil is considered healthy. That is carbon stored in the soil, or organic matter. Poor fertility, compaction, or salt accumulation are frequently indicated by pale, gray, or bleached soil. Yellow or reddish colors indicate mineral imbalances, drainage problems, or the presence of iron.

Another powerful indicator is smell. The scent of living soil is deep, sweet, and fresh, like the smell of earth after rain. The smell is that of microbiological activity. Anaerobic conditions, inadequate drainage, or high nitrogen inputs are indicated by aromas that are sour, musty, or ammonia-like.

Feel it now. Touch wet dirt with your fingertips. The texture of sandy soil is coarse and gritted. Clay has a dense, sticky feel. Silty soil has a flour-like consistency. The "tilth" of healthy soil is a soft, crumbly, flexible texture that retains its shape when squeezed but crumbles under light pressure. It should not crumble like dry dust or agglomerate into small bricks.

Make a little trench. Is it possible to see roots beneath the surface? Are earthworms afraid of light? Does water freely drain or pool? Compared to spreadsheet figures, these field indicators provide a more immediate image.

Through observation, you become more than just a soil manager—you become a soil steward.

Understanding Texture, Aggregates & Tilth

Hands and noses existed before lab findings and applications. One of a grower's most effective tools is still observation. There is no need for equipment; sight, smell, and touch can all provide information about the health, condition, and any issues of the soil.

Let us start with color. Black or dark, rich brown soil is considered healthy. That is carbon stored in the soil, or organic matter. Poor fertility, compaction, or salt accumulation are frequently indicated by pale, gray, or bleached soil. Yellow or reddish colors indicate mineral imbalances, drainage problems, or the presence of iron.

Another powerful indicator is smell. The scent of living soil is deep, sweet, and fresh, like the smell of earth after rain. The smell is that of microbiological activity. Anaerobic conditions, inadequate drainage, or high nitrogen inputs are indicated by aromas that are sour, musty, or ammonia-like.

Feel it now. Touch wet dirt with your fingertips. The texture of sandy soil is coarse and gritted. Clay has a dense, sticky feel. Silty soil has a flour-like consistency. The "tilth" of healthy soil is a soft, crumbly, flexible texture that retains its shape when squeezed but crumbles under light pressure. It should not crumble like dry dust or agglomerate into small bricks.

Make a little trench. Is it possible to see roots beneath the surface? Are earthworms afraid of light? Does water freely drain or pool? Compared to spreadsheet figures, these field indicators provide a more immediate image.

Through observation, you become more than just a soil manager—you become a soil steward.

Toolkit Box: Soil Test Interpreters and Apps

Deciphering soil test results can be overwhelming. Raw numbers don't mean much unless you know what to look for. That's where tools and interpreters come in.

1. Soil Balancing Calculators

Tools like the Soil Savvy platform or Kinsey-Albrecht charts help translate lab values into actionable insights. They show ideal ratios of calcium to magnesium, potassium to sodium, and how these influence structure and plant health.

2. SoilWeb and Web Soil Survey (USDA)

These government-backed tools help you understand the native soil type on your land. Input your location, and you'll see data on natural texture, drainage class, slope, and limitations. This helps you set realistic expectations and match crops to your native conditions.

3. Haney and PLFA Test Interpreters

For growers testing biological indicators, interpreting Haney or PLFA results requires guidance. Resources like the Bionutrient Food Association or Regen Ag Lab provide breakdowns of microbial biomass, carbon pools, and nutrient cycling.

4. Mobile Apps

Apps like Soil Mentor, Agrobase, and FarmOS help track soil tests, log observations, and provide visual graphs. Some apps integrate with remote sensors and allow you to store seasonal field notes. Think of them as your soil health journal.

5. Regenerative Coaches and Consultants

Sometimes a conversation is worth more than a chart. Soil consultants with a regenerative lens can help translate test data into management practices that fit your context. Look for advisors who work with biology, not just chemistry.

Don't test for the sake of data. Test to inform your next move. Good tools guide better actions.

Field Note: Diagnosing Clay with a Shovel and a Hose

Five raised beds on level clay ground made up the young, promising field. I had sown cover crops, stacked compost, and had faith that the soil would improve. However, things did not work out. Patchy germination occurred. The rate of drainage was slow. Plants had a hard time. I required clarification.

I thus took a shovel and a hose, as I should have done sooner.

I went straight to the first bed. The top three inches appeared respectable—loose, black, and root-filled. However, there was a thin, sticky, white coating behind it. Smear marks were left by the shovel. The hose's water gathered and sat. The lower layer did not disintegrate when I squeezed a handful of it; instead, it formed a compact ball. traditional clay behavior.

Although the surface had been enhanced by the compost, the underlying structure was still firmly in place. A pan had been formed by years of compaction that prevented water from escaping and roots from penetrating.

A change resulted from the diagnosis. I emphasized deep-rooted cover crops like sunn hemp and daikon radish, added extra fibrous organic matter, and used a broadfork to aerate. I began examining the entire soil profile instead of assuming compost would solve all of my problems.

The hose and shovel can sometimes speak louder than any lab ever can. They show the layers between what we think the soil needs and what it really needs, both physically and figuratively.

You must understand your soil as a layered, living, actual being rather than as an abstract idea in order to interact with it. You develop a relationship with the earth through testing, observing, and touching it. They enable you to see past the obvious and make plans of action that are in line with the realities of your soil.

You can identify what you are dealing with by the texture. Tilth and aggregates provide information about your biological health. You can detect where the soil is flourishing and where it needs assistance by using your senses of sight, smell, and touch. Apps and data provide context, but the reality comes from your hands, eyes, and shovel.

Avoid skipping the knowledge. It is the sole method for fostering growth. Soil only requires a partner who is attentive, not perfection. You will know what to do next if you know your soil. That is how you stop speculating and begin to gain self-assurance.

PART 2:
The Roots of the Revolution —
Foundations of Living Soil

Chapter 4:
Disturb It Less —
Transitioning to No-Till

Why Disturbance Matters

S omething is lost each time dirt is disturbed. A microbiological highway is destroyed. The months-long development of a fungal network is destroyed. A tunnel of worms comes to an end. Most significantly, the equilibrium between soil life, moisture, and air is disrupted.

Because soil is a dynamic, living organism, disturbance is important. When its structure is intact, roots may easily push lower, oxygen can enter, and water can enter. That natural architecture crumbles when disturbed, and we replace it with a temporary, loose façade that quickly compacts into dead slabs.

The loss of carbon is also accelerated by disturbed soil. When buried organic matter is exposed to air, it oxidizes quickly, generating carbon dioxide and reducing fertility over the long run. The rate of decrease increases with the amount of disruption. This jeopardizes the field's long-term productivity in addition to affecting plant health.

The foundation of the no-till philosophy is a mentality change: soil cannot be controlled by force. It is something to back by taking a back seat. allowing structure to be built by biology. allowing organic materials to accumulate. allowing worms, fungi, and roots to perform the tilling instead.

Tools for Small and Large-Scale No-Till

Going without tools is not the same as going no-till. It entails employing tools in a different way—carefully, precisely, and with an eye toward the future. The proper equipment makes it possible to apply no-till methods without upsetting structure or biology.

For Small-Scale Growers and Gardeners

1. Broadfork – Opens compacted soil without inversion. A foundational tool for permanent bed systems.

2. Wheel Hoe – Keeps weeds in check at the surface level with minimal disruption.

3. Silage Tarps – Prepares ground by smothering weeds and stimulating microbial breakdown of residues.

4. Compost Spreader – Distributes rich organic matter without digging.

5. Jang Seeder or Earthway Planter – Plants directly into compost or mulch layers with accuracy.

For Mid to Large-Scale Growers

1. Roller Crimper – Terminates cover crops by crimping stems and laying down biomass as mulch.

2. No-Till Drill – Plants seeds directly through residue into soil without tilling, preserving structure and cover.

3. Strip-Till Units – A low-disturbance option that tills only narrow planting strips while leaving the rest of the soil intact.

4. High-Residue Cultivators – Allow for shallow, residue-tolerant weed control.

5. Compost Spreaders and Manure Applicators – Key for nutrient management without inversion.

No-till requires efficiency, not absence of action. These tools allow growers to manage crops and weeds while respecting the underground ecology.

How to Go No-Till Without Losing Your Mind

Making the changeover to no-till is a process rather than a simple switch. Like individuals, fields require time to adapt. Be prepared for setbacks. Anticipate unsuccessful experiments. Expect surprises as well, such as deeper root systems, less erosion, improved moisture retention, and the first flush of mushrooms in compost mulch.

Choose a manageable plot first. If you are not prepared for the learning curve, do not try to do your entire farm at once. Concentrate on one area where you can try new things without jeopardizing your security or money.

Employ cover crops. These are your organic nitrogen fixers, weed suppressors, and tillers. Start with crops that grow quickly, such as field peas, oats, or buckwheat. Try mixtures with radish, rye, and clover throughout longer seasons. Leave the residue in place and roll or mow them down. The mulch of nature.

Spread mulch and compost. These will soften the soil, increase nutrient availability, and encourage microbial activity, just like tillage did. The plants are not being "fed" by you directly. The plants are fed by the earth, which you are feeding.

Use cardboard, tarps, or dense cover crops to suppress weeds. Early no-till requires effective weed control. When soil health improves over time, you will see fewer weeds.

Be patient. Awkward planting, difficult residues, and obstinate weeds can make the first year seem chaotic. However, by the second or third year, the soil starts to do the work for you. Before you realize how little work it took, you will be harvesting.

Maintain a journal. Keep track of the soil's reaction and what works and what doesn't. In a no-till system, observation takes the place of coercion.

DIY Corner: Build a Simple Compost Mulcher

Applying compost mulch is one of the finest strategies to create healthy no-till beds. Although you may buy completed compost, creating your own allows you to control quality and save money. Using a compost mulcher, you may smooth out planting surfaces and improve application by converting bulky compost into a finer layer.

What You'll Need:

- ☑ A large plastic barrel or metal drum
- ☑ A drill with a paddle or paint mixer attachment
- ☑ A mesh screen (½ inch hardware cloth)
- ☑ A few wooden planks or PVC pipes to build a simple frame
- ☑ Optional: a small electric motor with gear reduction for automation

Step 1: Prep the Drum

Cut the top off the drum and drill air holes on the sides to increase aeration while mixing. Mount the barrel on a wooden frame so it can rotate or spin in place.

Step 2: Add the Paddle

Attach a long paddle to your drill or motor shaft. This will act like a blender for your compost, breaking up clumps and mixing dry and moist materials.

Step 3: Screen the Output

Place a mesh screen over a wheelbarrow. After mixing, dump the compost onto the screen and sift out large sticks or unfinished material.

Step 4: Apply Immediately

This fine compost can be blended with your bed preparation or used as top-dress mulch. You will experience quick microbial colonization and consistent application.

This system pays for itself in a single growing season and costs less than $100 if you reuse parts. By feeding soil from the surface without ever using a shovel, mulching with compost is a silent revolution.

Try This On Your Plot: "Your First No-Till Bed in 3 Steps"

You don't need a perfect system to start. Here's a simple way to create your first no-till bed using accessible materials and biology.

Step 1: Smother and Prep

Draw a line across the width of your bed (3–4 feet, any length). Flatten or mow any vegetation. Spread a thick layer of newspaper or cardboard to block light, keep weeds at bay, and encourage worms to break up the soil.

Step 2: Add Layers

Cover the cardboard with 4-6 inches of completed compost. Add wood chips, grass clippings, leaf mold, or straw on top. This protects the soil, lowers evaporation, and provides food for bacteria, simulating the layering found on forest floors.

Optional: For added nutrition and biological diversity, scatter worm castings or rock dust on top.

Step 3: Plant

Using a hand trowel or dibble, make tiny holes in the mulch and compost. Plant seeds or seedlings straight into the hole. Take a deep breath. As they develop, the roots will penetrate the cardboard and investigate the softer soil beneath.

Keep an eye on moisture levels, cover the bed, and replenish mulch as necessary. You have just begun your first no-till plot, which means you will not be tilling, digging, or disturbing the soil.

Better moisture retention, stronger root penetration, and a soil that gradually turns into black, rich humus are all changes you will see right away. This has not been made available or useful yet. One bed at a time.

Trusting more—trusting the soil to function when it is not being pulled apart all the time—is the key to disturbing less. It is about allowing roots to spread out, allowing biology to recover, and giving oneself the gift of more gradual, steady, and sustained growth.

Making the switch to no-till can be challenging, particularly if you were raised believing that neat rows and bare soil are indicators of good health. However, no-till shows a deeper order that encourages diversity, thrives under mulch, and prioritizes soil life over surface management.

Begin modestly. Watch. Make deliberate tool selections. Make your beds carefully. You are voting for biology, abundance, and resilience with every inch you choose not to disrupt. No-till is proactive caring, not passive. Instead of fighting against the land, it is working with it.

Allow the earth to breathe. Let the fungus run. Allow worms to tunnel. Let the harvesting start. Watch it do more and disturb it less.

Chapter 5:
Disturb It Less —
Transitioning to No-Till

Organic, Synthetic, and Living Mulches

One of the most crucial no-till growing guidelines is to keep the soil covered. Vulnerable soil is bare soil. It loses carbon to the atmosphere, compacts readily, dries up rapidly, and attracts weeds. Soil is never left uncovered for very long by nature, and we should not either.

Mulching resembles the armor of nature. That translates to fallen branches and leaf litter in a forest. Mulch on a farm can take many different forms, each with its own advantages and disadvantages.

Organic mulches include straw, hay, shredded leaves, wood chips, compost, grass clippings, or even spent crops. These materials decompose over time, feeding the soil, improving tilth, and boosting microbial life. They're ideal for gardeners, market growers, and any operation focused on biological health. Compost mulch, in particular, creates a rich top layer teeming with life while suppressing weeds and holding moisture.

Synthetic mulches include plastic sheeting, landscape fabric, and biodegradable film. These materials provide more control over weeds and moisture but don't contribute to soil health. Plastic mulches can overheat the soil and must be removed eventually. Biodegradable films break down but may leave residue or uneven decomposition patterns. Synthetic mulches can work in some short-season cropping systems, especially in cool climates or weed-heavy fields.

Living mulches are crops planted to protect and nourish the soil while main crops grow. These could be low-growing clovers between rows, vetch under tall crops, or perennial ground covers in orchards. Living mulches offer erosion control, weed suppression, pollinator habitat, and nitrogen fixation—sometimes all at once. Their challenge is management: competition for water, space, or nutrients must be minimized with careful selection and timing.

In a no-till system, covering soil is not optional—it's essential. The right mulch improves moisture, feeds biology, and gives you more margin for success.

Cover Crop Basics as Living Mulch

The off-season is not the only time to use cover crops. Plants that protect the soil, cycle nutrients, and maintain biological activity are your living mulch. Consider cover crops as an integral component of your rotation strategy rather than as a respite between income crops.

In addition to working actively beneath the surface, a well-managed cover crop offers all the advantages of mulch. Their roots release sugars that support microbial life, loosen the soil, and support fungal networks. When properly terminated, their above-ground biomass turns into mulch in place, inhibits erosion, and shadows out weeds.

Popular cover crops include:

- **Cereal Rye** – Fast-growing, cold-tolerant, excellent for winter cover and weed suppression.
- **Hairy Vetch** – A nitrogen fixer that pairs well with rye in cool seasons.
- **Buckwheat** – Quick-growing summer option that flowers early and smothers weeds.
- **Daikon Radish** – A "bio-drill" with a deep taproot that breaks compacted layers.
- **Crimson Clover** – A gentle nitrogen fixer and pollinator attractor.
- **Sorghum-Sudangrass** – Tall biomass builder with natural weed suppression.

To use a cover crop as living mulch, timing is everything. Sow early, and let the crop develop enough canopy to shade weeds. Then manage it with a crimper, mower, or flail before planting into the residue. Some growers use relay planting—sowing the main crop directly into a maturing cover crop. Others terminate the cover fully and plant into the mulch.

Think of cover crops as dynamic mulch. They don't just sit on the surface—they grow, interact, and prepare the soil for whatever comes next.

Tarps, Smothering & Timing Tricks

The shortcut that many no-till producers are unaware they require is tarps. A straightforward silage tarp might revolutionize your soil preparation procedure without the need for tillage, herbicides, or arduous labor, despite the fact that it is not glamorous.

Silage tarps are long-lasting, UV-resistant plastic sheets that are either black or white on black. They provide a moisture-retaining, light-blocking surface when spread across a field or bed. Without light, weed seeds sprout and eventually wither. Under the tarp, plant materials and residue start to decompose more quickly, particularly when moisture is trapped. Because of the steady warmth and protection, soil life remains active.

To use tarps effectively:

1. Mow or crimp any existing cover.
2. Lay the tarp flat and weight the edges securely.
3. Leave in place for 2–6 weeks depending on the season and your goals.
4. Remove the tarp and plant directly into the softened, weed-free surface.

Timing is crucial. In spring, tarping beds early warms the soil, accelerates decomposition, and preps for transplanting. In fall, tarps clean up spent beds and protect them from winter erosion. In summer, tarps can smother aggressive weeds in fallow areas.

Tarps are a bridge—not a forever solution. They buy you time, suppress weeds without chemicals, and simplify bed transitions. When used strategically, they can turn your worst weed patch into your best growing space.

Common Grower Question: "Can I Just Use Cardboard?"

Of course. One of the easiest, most affordable, and most efficient tools for no-till cultivation is cardboard. When used correctly, it smothers sod, inhibits weed growth, and attracts earthworms to the area. It is widely accessible, biodegradable, and frequently free from nearby retailers or delivery services.

However, it must be used properly, just like any other tool.

When to use it:

- Converting a lawn or weedy area to no-till beds
- Establishing a new garden plot
- Lining pathways between beds
- Sheet mulching perennial zones or orchard alleys

How to use it:

- Remove any tape, staples, or glossy print. Only use clean, uncoated cardboard.
- Overlap pieces generously—at least 6 inches—to prevent light leaks.
- Wet the cardboard thoroughly to soften it and initiate decomposition.
- Add a layer of compost or rich organic material directly on top.
- Cover with mulch—straw, wood chips, or leaves—to insulate and weigh it down.

The cardboard decomposes and is eventually eaten by soil microorganisms. Roots will sift through. Fibers will be pulled deeper by worms. As the system establishes itself, it is your responsibility to maintain surface protection.

One word of caution: do not rely solely on cardboard. Although it works well for transition and weed control, active inputs like compost, cover crops, and biological diversity are necessary for long-term soil health. Consider cardboard to be an excellent place to start, not the entire system.

To safeguard life, soil must be covered. It is that easy—and that effective. Each mulch, tarp, and green cover serves as a barrier against erosion, a moisture buffer, a microbial feast, and a weed-control tactic all at once.

Organic mulches protect the soil and increase fertility. When utilized properly, synthetic mulches save time and accuracy. Living mulches infuse the system with vitality and biodiversity. Fallow season is transformed into fruitful time by cover crops. For bed transitions, tarps provide an effective, chemical-free solution. Furthermore, a worn-out lawn can be turned into a useful bed using simple cardboard.

Instead of ignoring issues, no-till farmers take steps to avoid them. You can give biology the advantage, conserve moisture, control temperature, and lessen weed pressure by regularly covering the soil. Better crops are the end outcome, but so is more resilient soil that gets stronger every season.

Begin where you are. Make use of what you have. Cover the earth. As you intentionally and carefully steer the system, let nature handle the heavy lifting. greater life will be protected and greater abundance will be found the more you cover.

Chapter 6:
Keep It Growing —
The Root Connection

Why Bare Soil Is Broken Soil

The lifeblood of a healthy soil ecosystem is living roots. Exudates, which are liquid carbon compounds, are released by each root tip and provide nourishment to the nearby microbial communities. These substances draw fungus and bacteria, which causes a surge in biological activity. In turn, that action stabilizes aggregates, cycles nutrients, and maintains the life-pulsing subterranean system as a whole.

Soil is functioning when there are roots present. Soil unravels in the absence of roots.

In addition to being unproductive, bare soil is biologically deformed. There are no exudates to sustain microbial life in the absence of plant cover. Nutrient cycles are inefficient without that microbial life. Carbon in soil decreases. Aggregates fall apart. Infiltration of water slows down. Erosion starts. Once-thriving subterranean life starts to deteriorate.

Weeds are opportunistic, yet they frequently rush in to fill that void. They do not develop the same advantageous connections with bacteria or fungi. They typically have short-lived, aggressive, and shallow roots. They take instead of giving.

Additionally, bare soil absorbs wind and solar heat. Changes in moisture and temperature can have an impact on microbial communities. Soils become too hot, too dry, or too unstable for long-term health when plants do not have a moderating effect.

It is critical to keep dirt covered. But it is important to keep it expanding. Soil biology is designed by roots. Without them, we are starving the system rather than feeding it.

Relay Planting, Interplanting, & Succession

The most efficient growers don't just plant—they layer. Relay planting, interplanting, and succession are strategies that keep roots in the ground continuously, maximizing biological activity, yields, and space.

Relay planting is the process of planting one crop before the previous crop has finished its cycle. For example, sowing bush beans between rows of garlic two weeks before the garlic is harvested. By the time the garlic is out, the beans are already established. This reduces downtime between plantings and ensures the soil is never idle.

Relay works well in no-till systems where disturbance is minimal. You avoid pulling out plants and disturbing the soil. Instead, you cut at the base and let the next crop grow into an already vibrant root zone.

Interplanting involves growing two or more crops simultaneously in the same space. Think lettuce under trellised tomatoes, or radishes alongside carrots. It mimics natural systems where multiple plant types coexist and share resources. The key is choosing crops with complementary growth habits—deep roots with shallow roots, tall plants with shade-tolerant companions.

This approach not only keeps the soil growing but also increases total productivity per square foot. More importantly, diverse root systems feed diverse microbial populations. You're not just harvesting more produce—you're cultivating complexity underground.

Succession planting is the art of planning continuous production. Rather than planting once per season, you stagger plantings to follow each other seamlessly. After early spinach, in goes summer squash. After the squash, a late-season cover crop or brassica.

In a no-till system, succession depends on gentle transitions. You cut crops at the base, mulch lightly, top-dress with compost, and plant again. You never till. The roots of the previous crop decay in place, feeding soil life and maintaining structure.

These techniques demand planning. But they reward you with higher yields, fewer weeds, richer soil, and continuous biological momentum. The more you keep the soil growing, the more it gives back.

Year-Round Crop Planning

One of the best strategies for soil regeneration is to maintain the productivity of your soil throughout the year. When no-till systems remain active, they flourish. This entails scheduling crops for fall, winter, and even shoulder seasons in addition to spring and summer.

Begin by charting the photoperiods, average temperatures, and dates of frost in your climate. These serve as a guidance for your planting windows. Certain crops, such as spinach or kale, can survive the winter in a variety of climates. Others require heat and full sun, such as corn or buckwheat.

Cool-loving vegetables such as radishes, arugula, peas, and lettuce are the most common in the spring. In order to make place for summer mainstays like tomatoes, peppers, beans, and cucumbers, they are finished early. As summer comes to an end, storage crops, beets, carrots, and brassicas take their place.

Winter should not be overlooked. You should be growing something even if you are not harvesting. Vetch, red clover, and winter rye are examples of cover crops that protect soil, increase biomass, and nourish subterranean life. Full winter harvests of turnips, leeks, carrots, and chard are achievable in milder climates. Low tunnels, often known as caterpillar tunnels, prolong the season in chilly climates.

A spreadsheet, a calendar, and a dedication to observation are necessary for this type of preparation. Timing crops involves more than just temperature; it also involves light, moisture, and soil warmth. It concerns whether anything will mature before frost or how quickly it grows after the solstice.

Year-round preparation has the advantage of keeping your soil healthy. You continue to have a link with the land, biology continues to function, and roots stay in the ground. Your harvests grow more consistent and durable as the work is spread out over the course of the year.

Above all, year-round planning changes the grower's perspective. You are not living in a boom-bust cycle anymore. You are fostering ecological, economic, and biological continuity.

Field Note: The Power of a Winter Rye Root

In a portion of a field that had been compacted for years, I planted cereal rye one winter. I had extra seed and an hour to spare, so I thought it could not hurt, thus it was not a deliberate repair. The rye appeared to disappear under the snow, grew to six inches, and sprouted before the first hard frost.

Spring arrived rainy and late. I checked that field slowly. I was taken aback when I finally did. The rye had blown up. The canopy was sturdy, windproof, and two feet high. But everything was altered by what was underneath.

I plucked one plant, and with it came the earth, dark and crumbly and alive. Long and branching, the roots wormed down farther than any shovel had ever dug in that plot. Water pooled less where the rye had been. Less evidence of erosion was present. And the results were unlike anything I had ever seen on that field before—vigorous growth, black leaves, and beds that were almost completely free of weeds—when I stopped it and planted squash.

In just five months, that winter rye root system accomplished more than ten years of tillage and a year of compost combined. It created a path for the roots of the following crop to follow, fed soil bacteria, and gently broke up compaction.

Roots are important. even when they are hidden from view. particularly when you are unable to see them.

Roots are where healthy soil begins, not the compost pile or fertilizer bag. Plants are more than just soil. They construct, fix, and power it. The lifelines connecting what transpires above and what changes below are their origins.

Every area of exposed dirt is a lost chance. A lost opportunity to cycle nutrients, produce structure, and support biology. The root connection is maintained throughout the year through succession, interplanting, and relay planting. They maintain the system's respiration, production, and motion. Planning for the entire year transforms a seasonal business into an ongoing partnership. It adds biological constancy and disrupts the stop-start cycle.

In addition to being anchors, roots also act as collaborators, communicators, and carbon conduits. When they are present, the earth flourishes. When they are not present, it decreases.

In no-till systems, the tilling, fertilizing, and structuring are all done by the roots. They establish the framework for plenty. As the grower, it is your responsibility to ensure that they never depart. Yes, keep it covered, but keep it growing more than anything else. Your harvest depends on your soil.

PART 3:
Feeding the Soil That Feeds You

Chapter 7:
Compost Systems that Actually Work

Cold vs. Hot Compost, Bokashi, and Vermiculture

Compost is transformation, not just rot. It is the process by which we transform garbage into a biological force. Compost has two functions in no-till systems: it provides nutrients to the soil and helps to create structure. However, not every compost system is created equal. Knowing how they differ can make the difference between fertile, flourishing soil and unsatisfactory outcomes.

Cold composting is the simplest, slowest method. It involves piling up organic matter—kitchen scraps, leaves, garden waste—and letting nature take its time. There is no regular watering, no turning, and no control of the carbon-to-nitrogen ratios. A cold pile may take 12 to 18 months to decompose if left unattended. For patient backyard growers who are not in a rush, it is perfect. Time and the sporadic survival of weed seeds are the trade-offs.

Hot composting is fast, intense, and highly effective. When properly prepared, a hot compost pile can achieve temperatures between 130 and 160°F, which kills weed seeds and pathogens. Regular rotation, appropriate moisture levels, and a balanced carbon-to-nitrogen ratio—ideally between 25:1 and 30:1—are all necessary. Manure and grass clippings give heat, while carbon is supplied by sawdust, straw, or shredded leaves. In as little as three to six weeks, a properly maintained hot pile can yield finished compost. For busy farmers who require a lot of compost all season long, it is ideal.

Bokashi composting is a fermentation-based system, not a decomposition process. It ferments kitchen scraps anaerobically using bran inoculated with beneficial microbes, typically lactic acid bacteria. Bokashi bran is used to tightly pack food waste, such as meat and dairy, into an airtight container. It turns into a pre-compost product in two weeks, which needs to be buried or added to a compost pile for ultimate decomposition. Bokashi is very nutrient-dense, compact, and odorless. Urban growers and those with limited space will find it extremely helpful.

Vermiculture, or worm composting, uses red wigglers to convert kitchen waste into castings—a concentrated, biologically rich soil amendment. Worm bins are easy to manage indoors or in shaded outdoor spots. The resulting worm castings are low-volume but potent, filled with enzymes, plant hormones, and beneficial microbes. Vermicompost shines in seed-starting mixes, transplant holes, and compost teas. It's ideal for anyone looking to add biological activity without bulk.

Each method suits different growers, climates, and scales. Choosing the right one means balancing time, space, labor, and compost needs. But whatever the method, the goal remains the same: feed the soil with biologically active, stable organic matter.

Compost Inoculants and Biological Boosts

Compost serves as a vehicle for life and is more than just nutrients. Bacteria, fungus, protozoa, nematodes, and actinomycetes are all abundant in high-quality compost. However, inoculants might give your pile a biological boost if it lacks diversity or colonizes slowly.

Inoculants are products or natural sources added to compost to enhance microbial diversity and speed up decomposition. These could include commercial blends of beneficial microorganisms, worm castings, fungal-rich forest soil, or completed compost. To feed microorganisms and hasten population growth, some growers incorporate humic acids, kelp, or molasses.

Completed compost is the most significant inoculant. A shovelful of high-quality, active compost adds billions of bacteria that are ready to work when it is added to a new pile. This is the pinnacle of microbial seeding. Worm castings are no different; a tiny amount of them yields remarkable biological diversity.

Fungal dominance can be encouraged by adding woody material, leaf mold, or fungal-rich soils from forest edges. Both bacteria and fungi are necessary for a healthy soil environment, but fungi are particularly important for nutrient cycling and aggregate formation in regenerative systems.

Compost teas, which are liquid extracts of compost biology prepared with oxygen and simple sugars, also contain inoculants. These teas improve disease resistance and biological communication by inoculating surfaces with beneficial organisms when sprayed on plants or soil.

Inoculants can take your compost from good to great, but they are not required for every pile. The key to fertility, particularly in places with monoculture or deteriorated soils, is biotic diversity.

How to Use Compost Without Overdoing It

Although compost has many benefits, more is not necessarily better. It is not a fertilizer; rather, it is a biological input. When used properly, it provides slow-release nutrients, promotes microbial life, and improves structure. Overuse can result in nutritional lockout, salt accumulation, and imbalances.

Compost works best in no-till systems as a living bed layer or as a top mulch. Consider ½ to 1 inch per bed, with leaf mulch or straw on top. This is similar to how organic stuff gradually builds up from the top of natural systems. It nourishes fungus that grow close to the surface without upsetting the soil's layers.

Additionally, you can apply compost only where you are planting, in bands or pockets. A scoop of compost placed right beneath the transplant offers a mild nutrient buffer for heavy feeders like tomatoes or squash without flooding the entire bed.

Do not thoroughly mix compost into the soil. Fungal networks in particular may be disrupted if the compost layer is inverted. Over time, let the microorganisms drag it down. Worms and other microorganisms will incorporate compost more effectively than any instrument in healthy soil.

Watch for signs of overuse. If plants show signs of excessive growth with poor fruiting, or if soil stays too wet and anaerobic, you may be applying too much. Some composts, especially those made from manure, can be high in phosphorus or salts.

Balance compost with carbon. Always follow compost with mulch or a living cover. This ensures microbial life stays protected and well-fed. Compost is not the whole system—it's a part of it. Respecting its role leads to long-term soil health, not short-term flash.

Compost Type	Best Use Case	Benefits	Considerations
Cold Compost	Mulch for perennials, pathways	Low-labor, slow-release nutrients	May contain weed seeds, slow to finish
Hot Compost	Annual beds, nutrient boost	Pathogen-free, quick turnaround	Requires management and space
Bokashi	Urban plots, food scrap recycling	Handles fats and meat, compact system	Requires burial or second-stage composting
Vermicompost	Seed starting, potting mixes	Extremely rich, biologically active	Small volumes, needs temperature control
Leaf Mold	Soil structure, fungal support	High fungal content, long breakdown	Seasonal availability
Manure Compost	High-demand crops, large beds	Strong nitrogen source, builds OM	Test for salts and pathogens
Mushroom Compost	Soil conditioner for flowers	Alkaline, softens heavy soils	Low nutrient content, high salts

Utilize compost according to your planting objectives and the requirements of your soil. Every type contributes special qualities. Over time, combining different compost sources creates a more varied ecosystem and lessens reliance on any one substance.

The silent powerhouse of wholesome no-till systems is compost. It creates structure that tillage could never duplicate, nourishes the soil food web, and turns waste into fertility. The compost pile is a home for life, a carbon sponge, and a biological reactor in addition to being a bin of rot.

The technique must work for you, whether you work slowly with cold compost, quickly with hot methods, compact with bokashi, or concentrate with worms. Progress, not perfection, is the aim. It is turning what would otherwise be thrown out into something useful. It involves learning to nourish the soil that sustains you.

Biological boosters and inoculants are tools, not panaceas. They assist you in amplifying the natural forces at work. However, they do not take the place of the necessity of restraint, balance, and observation. Compost is not a crutch; it is a support.

Compost becomes a concern when it is used excessively. It will continue to be beneficial if applied carefully, in the appropriate way, and at the appropriate moment. Use it as though you were creating something long-lasting rather than only repairing something temporary.

Composting is fundamentally an act of respect for soil, cycles, and decomposition. Closing the loop and maintaining local, rich, and regenerative fertility is one of the most straightforward and straightforward methods. Your soil performs well when your compost system is functioning properly. And everything else follows when your soil is healthy.

Chapter 8:
Cover Crops I — The Basics

Why Cover Crops Matter

Cover crops actively improve soil health, resilience, and productivity rather than only serving as off-season fillers. Cover crops are a great addition to any no-till system, whether it is being used on a hundred-acre field or in a backyard. Numerous vital tasks are carried out by these unsung heroes, including preventing soil erosion, adding organic matter, fixing nitrogen, controlling pests and weeds, creating root channels, promoting microbial activity, and even releasing nutrients that would otherwise be trapped in the soil.

The fact that cover crops serve as live roots is among their most underappreciated uses. Degradation of bare soil happens rapidly. It loses its microbiological vitality, is compressed, and loses moisture. Cover crop roots release sugars and other substances that provide food for worms, bacteria, fungus, and protozoa. By forming aggregates, these microorganisms enhance soil structure, liberate nutrients that have been trapped in minerals, and establish long-term carbon stores. The soil practically comes alive with cover crops.

Additionally, cover crops act as organic mulch both during and after growth. Weeds are kept from receiving light by a dense stand of vetch or rye. After being sliced or crimped, the residue gradually feeds the soil below while still inhibiting weed germination. They enable the grower to disrupt pest cycles, cut down on inputs, and protect against drought and high heat. In every aspect, cover crops are an essential component of regenerative no-till farming systems and are neither a luxury nor a side project.

Timing and Bed Prep

Timing is key if you want cover crops to function for your system. Weak growth, inadequate coverage, and diminished advantages are the outcomes of planting too late. Planting too early can cause issues with rotation or interfere with the harvest of your cash crop. Selecting the appropriate time is aided by knowing your crop succession plan, soil temperature windows, and frost dates.

The objective is to grow cover crops in the fall as soon as summer crops are removed. This could entail redoing a bed rapidly following a late-summer planting or overseeding into standing crops a few weeks prior to harvest. Cool-season cover crops such as hairy vetch, Austrian winter peas, or cereal rye should ideally be planted at least four to six weeks prior to the first hard frost. Even eight to ten weeks of growth may be feasible in warmer climates, enabling complete canopy closure and optimal biomass.

Timing is more critical for spring cover crops. As soon as the soil is workable, it is time to plant early-season coverings such as phacelia, field peas, or oats. That could result in windows melting in colder climes. Before relocating heat-loving summer crops, February or March seeding might provide a burst of growth in milder areas.

The way you prepare your bed depends on how your soil is doing right now. It could be necessary to cover existing vegetation with a tarp when using a new no-till system, and then plant the cover crop straight into the exposed area. Light raking or broadforking prior to sowing enhances seed-to-soil contact in established no-till beds. Broadcast seed can be anchored and given an early nutrient boost by lightly sprinkling it with compost.

After seeding, consistent wetness is crucial. For seeds to sprout, even drought-tolerant coverings require moisture. Timed rainfall windows or drip irrigation can be beneficial. Enhancing emergence can also be achieved by walking the seed in or pressing it in with a roller.

A bed that maximizes growth potential and reduces disruption is the finest prepared bed.

Seasonal and Regional Recommendations

Choosing the right cover crop depends largely on when you're planting and where you're growing. Not every species will thrive in every climate or season. Success comes from matching your cover crop to your region's rainfall, temperatures, and soil conditions.

Cool Season (Fall to Spring)

- *Cereal Rye:* The workhorse of fall covers. Cold-hardy, weed suppressing, and deep-rooted. Best in northern climates.

- *Hairy Vetch:* A nitrogen fixer that partners well with rye. Needs time to establish. Great biomass producer.

- *Winter Peas:* Another nitrogen fixer. Less winter-hardy than vetch but good for mild climates.

- *Crimson Clover:* Adds nitrogen and attracts pollinators. Overwinters well in moderate zones.

- *Tillage Radish:* Breaks compacted layers with a long taproot. Dies off in cold winters, making spring management easy.

Warm Season (Spring to Early Fall)

1. *Buckwheat:* Fast-growing, great for weed suppression. Perfect between spring and summer crops.

2. *Sorghum-Sudangrass:* Excellent for biomass, erosion control, and weed management. Requires warm soil.

3. *Cowpeas:* Fixes nitrogen, tolerates heat and drought. Suitable for southern climates.

4. *Sunflowers:* Build structure, attract pollinators, and tap into deep nutrients.

5. *Soybeans (as cover):* Provide nitrogen and canopy quickly in summer systems.

Perennial or Multi-Season Options

- *White Clover:* Long-term living mulch for orchards or pathways. Fixes nitrogen, tolerates mowing.

- *Alfalfa:* Deep-rooted, excellent nutrient scavenger. Works in pastures or rotational fields.

Your decision should be based on the normal rainfall and dates that are frost-free in your area. Buckwheat and cowpeas are examples of quick biomass builders that grow well in humid southeast regions. Sorghum and tillage radish are examples of deep-rooted crops that thrive in arid high-desert environments. Cereal rye is frequently the only practical fall-planted crop that endures until April in northern regions with lengthy winters.

Known as a "cover crop cocktail," a mixed-species approach adds resistance. Different microorganisms are fed by different roots. Some scavenge nutrients, while others fix nitrogen. Some cover quickly, while others continue into the following season. In fluctuating situations, buckwheat and cowpeas or cereal rye and vetch might be mixed to improve overall performance.

A cover crop that is appropriate for your area helps buffer your soil biology, lessen erosion, and prepare the ground for the next crop to thrive, regardless of the season.

Try This On Your Plot: "Three Easy Starters"

If you've never planted a cover crop before, start small. Here are three simple, forgiving, and effective cover crop strategies for new and experienced growers alike.

1. Fall Rye and Vetch Mix

- *When to plant:* 4–6 weeks before first frost

- *Where it works:* All climates, especially cold zones

- *Why it works:* Rye suppresses weeds and scavenges nitrogen. Vetch fixes nitrogen and adds biomass. Together, they offer full soil coverage and build rich organic matter.

- *How to plant:* Broadcast 60–80 lbs/acre of rye and 20–30 lbs/acre of vetch. Water thoroughly. Terminate in spring with roller-crimper or by cutting at bloom.

2. Summer Buckwheat Blitz

- *When to plant:* Late spring to midsummer

- *Where it works:* Most regions, especially for weed-choked beds

- *Why it works:* Buckwheat germinates in 3–5 days, shades weeds fast, and builds soft biomass. Easy to terminate and great for short windows between crops.

- *How to plant:* Broadcast 40–60 lbs/acre. Rake lightly or dust with compost. Mow at flowering to prevent reseeding.

3. Oats and Field Peas for Spring Beds

- *When to plant:* Early spring, as soon as soil can be worked

- *Where it works:* Cool climates with wet springs

- *Why it works:* Oats provide quick cover and structure. Field peas fix nitrogen and build root networks. Both die back in frost or can be cut easily.

- *How to plant:* Drill or broadcast a mix at 80–100 lbs/acre. Plant summer crops into residue or mulch over.

Just one or two beds should be used for these. Observe how the weeds act, how the earth reacts, and how the moisture retains. Maintain a journal or photo log to document the changes. These three combinations will help you understand how cover crops change the fertility and energy of your growing area in a practical way.

Although cover cropping is simple, it has great power. After witnessing the transformative power of living roots and green biomass, you will never again leave a bed naked.

In a regenerative no-till system, cover crops are essential and cannot be ignored. While you are not looking, they repair the harm caused by extraction, develop the soil, and create the foundation for biological vigor, resilience, and productivity. You are making an investment in future harvests with each cover crop you plant. Each root in the earth is a biological greeting to the ecosystem of your soil.

The fundamentals are simple: match the crop to your objectives, the land, and the season. Plant at the right time, carefully prepare your beds, and let nature take care of the rest. It is not necessary to become an expert at cover cropping all at once. Begin with a single bed. Watch. Modify. Give it another go.

As you incorporate additional cover crops into your system, your soil will start to self-regulate, self-structure, and self-heal, reducing your need for inputs. Instead of being regulated, the best productive soil is promoted. Furthermore, one of the best strategies to promote life's growth from the ground up is to plant cover crops.

Plant them. Take a lesson from them. Additionally, allow them to train your soil how to flourish once more.

Chapter 9:
Cover Crops II —
Advanced Techniques

Multi-Species Cocktails

There is nothing gimmick about a cover crop cocktail. It is a thoughtfully curated combination of plant species that collaborate to create intricacy beneath the surface. This method is similar to the complex, layered, and interconnected ecosystems found in nature. In a monoculture, microbial stimulation and a limited band of root exudates are applied to the soil. It gets a whole menu in a cocktail. The variant is advantageous to nematodes, earthworms, fungus, and bacteria. A patchwork of biological support is produced by variations in root depth, growth patterns, and seasonal activities.

A simple cocktail could include daikon radish, hairy vetch, and cereal rye. Rye disrupts disease cycles, inhibits weed growth, and increases aboveground biomass. Vetch fixes nitrogen and leaves behind a protein-rich residue that feeds microbes. Radish scavenges deep minerals and breaks up compacted layers. They are significantly more beneficial when combined than when used separately.

Blends that are more complex may contain ten or more species. Structure is provided by small grains such as barley, triticale, or oats. Field peas, fava beans, and crimson clover are examples of legumes that fix nitrogen. Pests and diseases are suppressed by brassicas, such as mustard or kale. Broadleaves, such as flax or sunflower, nourish fungal networks and provide support to beneficial insects.

Understanding function rather than just species count is essential to a cocktail's success. Choose plants that serve a variety of ecological functions. Combine fibrous roots with taproots. Combine nitrogen fixers and users. Combine slow-builders and fast-growers. Consider it the balance of a subsurface living economy.

Making adjustments is necessary when planting cocktails. The easiest approach is broadcasting, which is followed by composting or light raking. For larger plots, row planting or drilling may be more effective. It is natural to expect uneven emergence. Uniformity is not the aim. It is long-term biological enrichment, resilience, and diversity.

There is more to cocktails than biomass. They have to do with relationships. More life equals stronger, more self-sustaining soil, and each new species attracts more life.

Termination Without Tilling

While they are growing, cover crops are helpful, but after they are terminated, they alter. At that point, their leftovers begin to nourish the soil, suppress weeds, and form a layer of mulch for protection. Termination in a no-till system needs to take place without disturbing the soil. Plowing, inverting, and ripping are all prohibited. A number of efficient methods transform cover crops into soil-building tools while maintaining the soil's structure.

Crimping is one of the most common no-till termination methods. A roller crimper—whether tractor-mounted or walk-behind—lays down the cover crop and crushes the stems, halting growth. Timing is essential. Crimp too early, and the plant may regrow. Crimp at flowering, and you ensure a clean kill. Rye and vetch crimp well when fully headed or in bloom. Buckwheat, oats, and radish are best terminated before seeding to avoid volunteers.

Cutting is another simple method. A flail mower, weed trimmer, or sharp scythe can chop down the cover crop and leave residue on the surface. Cut high to preserve soil cover. Avoid chopping material too finely—larger pieces decay more slowly and provide longer-lasting mulch.

Tarping is ideal for small-scale or garden operations. After mowing or crimping, lay a silage tarp or UV-blocking plastic over the bed for two to four weeks. The darkness prevents regrowth and speeds decomposition. When the tarp is removed, the bed is soft, weed-free, and ready for planting.

Winterkill is a passive strategy for cold climates. Choose cover crops that die off in freezing weather, such as oats, buckwheat, or tillage radish. Their decayed roots improve structure and drainage, and their residues break down without intervention.

The goal with all termination methods is to stop growth while preserving root channels, protecting the soil surface, and supporting microbial activity. No-till growers build with biology, not blades. And proper termination is where the biology really starts to take over.

Cover Crops in Rotation with Cash Crops

You do not have to give up income crops to incorporate cover crops into your rotation. It entails matching the long-term health of your soil with your crop plan. The greatest growers actually view cover crops as a component of the rotation rather than as something distinct. They are just as necessary as any grain or vegetable they grow.

A straightforward two-part rotation may include a fall rye/vetch mix after spring-planted crops. A summer crop of tomatoes is produced when the rye is finished growing in the spring after growing over the winter. The cover crop acts as a bridge in this system, restoring and safeguarding the soil in between harvests.

In more intricate systems, winter cover crops and summer cash crops rotate, and occasionally a fast-growing summer cover, such as buckwheat, is inserted in between spring and fall crops. Permanent cover crops can serve as live mulches in orchards and vineyards all year round, and grazing or mowing can be utilized as management techniques.

Relay cropping integrates cover crops into the standing cash crop. For example, seeding crimson clover under a canopy of sweet corn a few weeks before harvest. The cover crop establishes slowly, gets full light after harvest, and takes over without soil disturbance.

Interseeding can occur between rows of established cash crops. In a no-till cornfield, for instance, growers may broadcast rye and clover into standing rows at the V5 stage, ensuring ground cover takes over after corn senescence.

When given the same consideration as cash crops, cover crops perform best in rotation. Take into account when they should be terminated, how they affect subsequent crops, how they manage residue, and how they affect roots. Only when they are timed to fit neatly into your system may cover crops help you achieve your highest yields.

Consider them an investment rather than a disruption, one that will pay off in the form of healthier crops overall, reduced input costs, improved water retention, and fewer weeds.

Of course. Actually, "cut and drop" is one of the most tasteful, approachable, and ecologically responsible methods of managing cover crops in garden and small-scale systems. It entails chopping off the base of the cover crop while leaving the biomass in place as a layer of mulch.

This method replicates how plants naturally die, fall, and nourish the soil on a forest floor without ever being taken out. The roots remain in the ground, sustaining fungi and microorganisms. The tops stay on the surface, keeping moisture in, inhibiting weed growth, and gradually breaking down into nutrient-rich material.

- **When it works best:** The best crops for cut and drop are those with soft stems, such as clover, oats, or buckwheat. These plants decompose rapidly and do not create thick mats that obstruct planting. Although their fibrous waste takes longer to decompose, hairy vetch and rye can also be used in this manner.

- **How to do it:** Cut the cover crop at the base using a sharp knife, sickle, or hand shears. Evenly distribute the cuttings around the bed or in between rows. If necessary, you can weigh down the mulch or add a thin layer of compost on top to speed up decomposition.

- **Benefits:** This method saves effort, avoids using machinery, and protects soil life. It is particularly helpful on small plots, raised beds, and urban gardens where crimpers and mowers are impractical. With no engines, no pollutants, and no soil turning, it is a silent method.

- **Limitations:** Cut and drop can host slugs or slow down planting in humid climates. Without anchoring mulch, the residue may move in high-wind zones. Additionally, as they break down, some high-carbon coverings, like rye, may momentarily bind nitrogen.

Cut and drop is still a mainstay for many producers, particularly those who are adopting hand-scale no-till. It enriches the soil and streamlines the system.

Leave your cover crop in its natural habitat. Give it to the microorganisms that gave it food. Allow it to finish the loop.

A simple soil-building method is transformed into a dynamic system that feeds itself, fixes itself, and becomes better over time with the help of advanced cover crop techniques. Cocktails of different species optimize resilience and variety. No-till termination maintains biological activity and structural integrity. With cash crops, strategic rotation guarantees that no soil is ever left behind. Additionally, straightforward techniques like cut and drop return the control of soil management to the producer.

The more cover crops you use, the more you notice their impact on your thinking about your land as well as in the field. One root at a time, they create long-term health, slow down the pace, and lessen the noise.

Complicated does not equate to advanced. It denotes intentionality. It entails considering each plant, season, and soil interaction as a component of the total.

Gaining proficiency in these methods places you at the center of a dynamic, expanding, and robust system. Soil is never idle in this place. One in which health improves, inputs decrease, and the earth begins to work for you rather than against you.

Let the coverings take care of themselves. Allow the roots to grow deeply. Allow the soil to recall what life is all about.

Chapter 10:
Fertility Without Chemicals

Soil Microbiology as the Fertility Engine

A package of fertilizer is not the first step toward fertility. Life in the soil is the first step. The biological basis for how nutrients are transported, unlocked, and shared with plants is formed by that life, which includes bacteria, fungus, protozoa, nematodes, earthworms, arthropods, and more. The natural fertility engine that drives each effective no-till system is produced by this complex population of soil organisms.

Microbes do not play in the background passively. They actively fix nitrogen in the atmosphere, break down organic materials, exchange nutrients with plant roots, and collect nutrients from minerals. By establishing networks between several plants, mycorrhizal fungi improve water efficiency and phosphate uptake. On the roots of legumes, nitrogen-fixing bacteria develop nodules that transform inert air nitrogen into a form that may be used. Crop leftovers, leaves, and compost are broken down by decomposer fungi and bacteria to produce stable humus and accessible nutrients.

Fertility becomes self-regulating when this biology is flourishing. In accordance with plant requirement, nutrients are delivered gradually. Stabilized and stored organic materials. Resistance to pests increases. Root zones remain active for a longer period of time. You are creating a cooperative system that more accurately satisfies their demands than any synthetic input, as opposed to force-feeding crops.

Although chemical fertilizers can produce benefits right away, they bypass the biological cycle. They make plants nutrient-dependent, inhibit fungus, and decrease microbial diversity. This eventually results in compaction, nutrient runoff, fragile soil structure, and decreasing yields. The soil ceases to be living and turns into a conduit for inputs.

Fertility in a no-till setting is renewable due to biology. It is more than just the amendments, minerals, or compost. The cycle is propelled forward by life.

Compost Teas, Extracts & Amendments

The soil biology requires fuel once it is established. A key component of any no-till system is compost, but how you utilize it makes all the difference. In addition to applying compost in bulk, liquid compost forms, such as teas and extracts, provide a means of introducing specific biology into plants and soil.

Compost teas are brewed by soaking high-quality compost in aerated water, often with small amounts of molasses, kelp, or fish hydrolysate. The brew typically runs 12 to 48 hours with active bubbling to keep oxygen levels high and microbes multiplying. The result is a living liquid that can be applied to soil or sprayed on foliage to introduce beneficial bacteria, fungi, and protozoa. It's a biological inoculant that strengthens root zones, suppresses pathogens, and improves nutrient cycling.

Compost extracts are simpler—no brewing, just soaking and agitation. You place compost in a mesh bag, swish it in water for 10–20 minutes, and apply the liquid directly. Extracts contain fewer microbes than tea but are quicker to prepare and excellent for soil drenching.

Biological amendments extend fertility without chemicals. Worm castings, humic acids, sea minerals, and fermented plant extracts add microbial diversity, micronutrients, and chelating agents. They don't force growth—they support the soil's natural systems so that growth happens more efficiently.

These inputs are not fertilizers in the conventional sense. They don't spike nitrogen levels or deliver instant green-up. What they do is support the ecology of the soil, helping your system build fertility from within. With every dose of compost tea or worm extract, you're adding memory, resilience, and health to the soil—not just a nutrient.

Balancing Nutrients Ecologically

Creating interactions between components that complement the structure and biology is more important than simply adding minerals to the soil in order to achieve balanced fertility. Fungal growth can be inhibited by an excess of nitrogen. Too much phosphorus can obstruct zinc. Compaction or crusting may result from an imbalance between calcium and magnesium. Synergy, not saturation, is the aim.

An adequate soil test, ideally one that provides information on micronutrients, base saturation, and organic matter in addition to NPK, is the first step towards ecological nutrition balancing. It is necessary to move from correction to cooperation in order to interpret these findings.

Calcium supports structure and root development. It flocculates soil particles, improving tilth and drainage. But too much calcium relative to magnesium can dry out sandy soils. Balance matters more than volume.

Magnesium holds water and tightens soil. In clay-heavy soils, high magnesium can lead to compaction and poor drainage. Bringing in gypsum or soft rock phosphate may help balance the effect without disturbing the system.

Potassium supports plant strength, carbohydrate transport, and disease resistance. It leaches easily from light soils but can build up dangerously in manured fields or with excessive compost.

Micronutrients like boron, zinc, copper, manganese, and molybdenum play critical roles in enzyme activation and plant metabolism. They're needed in small amounts but should not be ignored. Often, a deficiency of one micronutrient limits the entire system's performance.

Using soft supplements like rock dust, kelp, sea minerals, basalt, charcoal, and high-diversity compost, ecological fertility raises background levels instead of using artificial fixes to address shortages. Slowly and without upsetting the system, these inputs nourish the soil as well as the plant.

Repeated testing and adjustments are not necessary for a healthy no-till soil. With balanced inputs and biological activity, it gradually increases fertility. This fertility is stored not only in soluble nutrients but also in microbial biomass and organic materials. On a chart, it becomes more than just numbers. In a biological system, it turns into resilience.

Infographic: Nutrient Flow in a No-Till System

To better visualize how nutrients move in a healthy no-till system, imagine a loop—a closed cycle driven by roots, microbes, and surface organic matter:

1. Roots release exudates.

   Sugars, amino acids, and organic acids are secreted by roots. These feed soil microbes and stimulate fungal growth.

2. Microbial life activates nutrients.

   Bacteria decompose organic matter. Fungi scavenge phosphorus. Protozoa consume bacteria and release plant-available nitrogen. Mycorrhizal fungi extend the root zone and help absorb nutrients.

3. Organic matter breaks down.

   Surface mulch and compost decompose slowly, releasing a steady supply of nutrients into the root zone. Residues are processed by worms, beetles, and microbes.

4. Nutrients move through the plant.

   As nutrients are absorbed, they're used for growth, reproduction, and defense. Surplus minerals and carbon compounds are returned to the soil through leaf fall, root die-off, and crop residues.

5. Microbes store and cycle nutrients.

   Microbial biomass holds nutrients in a stable form until released by predation or decomposition. This acts as a buffer against leaching and overapplication.

6. Amendments enhance, not override.

   Compost teas, extracts, and ecological minerals are introduced in small doses to stimulate microbial activity, fill gaps, and support overall balance.

Nothing is wasted in this loop. Each input serves as a tool to preserve long-term fertility and fortify the biology. It is a dynamic network of life, decay, release, and rebirth rather than a linear system of extraction.

Without the use of pesticides, fertility is a reality based on biology, diversity, and the grower's perseverance. Fertility is a natural consequence of treating soil as a living, breathing partner rather than as an inert medium.

The fertility engine in a no-till system is microbiological. Compost, root exudates, and close observation are its main sources of energy. Living strength is added via compost teas and extracts. Balance without exhaustion is supported by soft modifications. The cycling of nutrients is first biological, then chemical. Although there are no short cuts, there is depth—depth in biology, comprehension, and outcomes.

You nourish the soil, not the plant. The plant is fed by the soil. It is a change in perspective and approach. It has nothing to do with your application. It all comes down to what you support. Slow, steady, and cumulative is how real fertility works. The source is not a jug. It expands in tandem with the system.

Soil builders, not soil managers, are the best growers. You have to foster fertility rather than purchase it. You are creating the groundwork for an autonomous future with each cover crop, compost layer, and microbial input. That is long-lasting fertility. That is chemical-free fertility.

Chapter 11:
Water and Soil —
Build a Sponge, Not a Sieve

Soil Moisture Retention Strategies

Similar to a sponge, soil should retain water during rainy seasons and gradually release it to plant roots during dry ones. Sadly, a lot of traditional growth methods make soil into a sieve; water seeps through, evaporates rapidly, and leaves the ground compacted, dry, and in need of continuous irrigation. No-till farmers take a different approach to soil. By enhancing biological function and structure from inside, rather than using external technology, they hope to boost water-holding capacity.

Structure is the first step in moisture retention. Moisture is retained in the organic matter and in the pores between the particles of well-aggregated soil. By retaining water close to root zones and slowing down evaporation, these aggregates function as tiny reservoirs. These aggregates are broken up by tillage, which also lessens the soil's capacity to hold onto moisture like a sponge. On the other hand, no-till systems promote better penetration and longer moisture retention by allowing aggregates to form and consolidate.

Root mass is another important tactic. In addition to providing nourishment for plants, roots also leave behind organic debris, channels, and root exudates that support microbial populations. These microorganisms promote water storage, increase porosity, and bind soil particles. Additionally, dead roots produce microscopic spaces that function similarly to small sponges. Your soil can hold more moisture if your system has more roots, particularly from cover crops.

Surface cover is just as important. In the sun, bare soil loses water rapidly and bakes. A layer of mulch, such as agricultural debris, compost, straw, or leaf mold, protects the soil from the sun and wind and lowers evaporation. Additionally, by slowing down rainfall, this layer permits water to seep in rather than wash off.

Building resilience is the goal of water retention, not merely preserving irrigation. Strong biologically structured soils perform better in dry spells than extensively irrigated but dead farms. They cling to and extend what nature provides them.

Irrigation, Swales, and Mulch Layers

In many climates, irrigation is still required even in soils that are biologically rich. However, there is a significant variation in the way water is provided and kept. Minimal disruption is included by the no-till philosophy, and water systems ought to be considerate, effective, and a part of the living system.

Drip irrigation is often favored in no-till systems. It delivers water directly to the root zone, minimizing evaporation and keeping the soil surface dry, which helps control fungal issues and weeds. Lines can be laid under mulch or compost, creating a protected zone that remains moist and biologically active. Drip is especially effective for row crops, raised beds, and perennial systems where water can be targeted precisely.

Overhead watering, such as sprinklers or pivot systems, has advantages in some contexts. It can mimic rainfall, evenly saturate large areas, and help germinate small-seeded crops. However, it also wets the foliage, increases disease pressure, and loses water to evaporation. On bare soil, overhead watering can cause crusting, runoff, and compaction. When used in conjunction with mulch or living cover, though, it can still be effective and beneficial.

In addition to irrigation systems, landscape design is important. Rainwater is captured and infiltrated by swales, which are level, shallow ditches excavated along contour lines. Water is absorbed and retained in the root zone rather than flowing off the ground. Swales are particularly effective on farms that want to capture water rather than drain it, on hillsides, and in perennial orchards. Swales can be transformed into permanent fertility trenches by mulching them with woody waste, compost, or leaves.

Mulch layers are your passive irrigation system. Straw, wood chips, cover crop residues, and even woven fabric create a microclimate that reduces evaporation, buffers temperature swings, and prevents splash erosion. A properly mulched bed can go days or even weeks longer between waterings compared to bare soil.

Water isn't just about delivery—it's about distribution, infiltration, and conservation. When your soil becomes a sponge, every drop counts more.

Drought-Proofing Through Organic Matter

The most effective item in a grower's toolbox for controlling water is organic matter. Root biomass, compost, and cover crop residue all help to improve the soil's capacity to hold water. An extra 20,000 gallons of water can be stored per acre for every 1% increase in organic matter. That is a measurable, observable fact, not a theory.

Organic materials have a reservoir-like effect. Rain or irrigation causes it to absorb moisture, which it then gradually releases when the soil dries out. Additionally, it provides nourishment for the microbial communities that form aggregates, which expand pore space and lessen compaction. Tight clods or firm crusts are not formed by soils that contain a lot of organic matter. They remain pliable, supple, and permeable.

Compost, aged manure, and mulch are the first inputs needed to build organic matter. However, it does not end there. Long-term carbon benefits from living roots are equal to or greater than those from other sources. Sugars released by each root support microbial life, which in turn creates exudates and glues that bond particles and create humus.

This structure is preserved by avoiding tillage. Tillage speeds up the oxidation and breakdown of the soil while also temporarily incorporating organic substances. Long-term consequences are loss rather than gain. Conversely, no-till systems permit organic matter to build up on the surface, where it will eventually be pulled downward by microorganisms and worms.

The key to drought-proofing your soil is not new technology, but rather ancient biology. The demand for external irrigation drastically decreases when soil can hold water, protect against extremes, and feed itself. Your reliance on weather forecasts decreases, your harvests hold steady throughout dry spells, and your plants keep greener for longer.

Your soil can hold more water the more carbon you add and protect.

Like many things in no-till gardening, the answer to this frequently asked topic depends on your system, your crops, and your goals. Overhead and drip both offer benefits and limitations. Matching the approach to the situation is more important than making a rigid choice.

The effectiveness of drip irrigation is exceptional. It provides water to the roots, where it is most required. Since the aisles and pathways remain dry, it inhibits the growth of weeds. It works with tarp systems as well as mulch and compost. Drip is used for high-value row crops including melons, tomatoes, and peppers. Perennials and permanent beds both benefit greatly from it.

The disadvantages? Drip lines need to be installed and maintained. Lines may move, emitters may clog, and rodents may chew through pipe. Planning is necessary to control homogeneity and pressure in large systems. Furthermore, drip finds it difficult to cover wide, open spaces. It works best in settings that are structured and closely monitored.

Rain is simulated by overhead irrigation. It swiftly and uniformly covers wide areas. It works wonders for seed germination and crops like lettuce, spinach, and direct-seeded greens that require steady top-down moisture. In smaller plots, overhead systems are relatively economical and easy to install.

However, in humid climates, overhead may also encourage disease, increase evaporation, and water weeds. It can cause runoff and pooling on soils that are compacted or prone to crust. However, such drawbacks can be lessened when combined with mulch and a solid soil structure.

Is drip always preferable, then? Not all the time. It works well in arid areas with long-term plantings, high-value crops, and structured beds. Overhead can be used as a temporary fix during crop establishment, in huge areas, or with cool-season crops.

The biology is more important than the method in a really resilient system. Watering is but one aspect of the problem. The remaining factor is how effectively your soil retains and releases that water when it matters.

Water is a rhythm as well as a resource. It passes through soil pores, microorganisms, and roots as a living component of your growth system. When the soil is treated like a living sponge, it absorbs and retains that rhythm before gradually releasing it where and when it is most required.

Water is not the only thing no-till farmers do. They increase the soil's ability to efficiently use and store water. They keep every drop in play, prevent evaporation, and prevent runoff. They turn dry soil into a robust, moisture-retaining soil by using organic matter, mulch layers, root biomass, and organized systems.

Smarter irrigation is achieved. The mulch gets deeper. Roots extend farther. Organisms in soil put in more effort. And rather than working against nature, the farmer works with it.

Gear and gadgets are not the focus of this chapter. It involves the steady, silent process of making soil that does not freak out during a dry week. The goal is to create a buffer. A margin. A backup.

As with a sponge, cultivate soil. Give it food. Put a lid on it. Give it what it requires. Because the roots are where the best irrigation system begins. And the soil is where the ideal water management plan starts.

PART 4:

Grow Anywhere —
Adapting to Any Plot

Chapter 12:
No-Till in Urban, Suburban, and Rural Settings

Growing in Containers, Beds, Yards, and Acres

No-till farming is not constrained by area or location. The fundamentals of soil health—minimum disturbance, continuous cover, and active biology—apply whether you are maintaining a rural farm, an urban lot, or a suburban backyard. The scale and the approach are what shift. It is possible to cultivate soil and produce plenty in any environment.

In urban settings, space is limited, but opportunity abounds. Sidewalk strips, rooftop planters, balconies, and raised beds offer more potential than many realize. Here, containers play a vital role. Large grow bags, wooden boxes, upcycled barrels, and fabric pots become your soil systems. Fill them with living compost-rich media, inoculate them with worm castings or compost teas, and top with mulch. These micro-ecosystems respond to the same no-till principles: don't dig, keep them covered, and feed them with organic matter.

For those with access to land in a city—community gardens, vacant lots, or schoolyards—permanent raised beds can serve as intensive no-till hubs. Constructed with wood, stone, or even sandbags, these beds reduce compaction, manage runoff, and keep inputs targeted. Use compost and mulch instead of tilling, and plant densely to shade out weeds. Urban no-till thrives on precision, creativity, and building soil above degraded ground.

Suburban yards offer unique opportunities. These spaces often contain patches of lawn, ornamental beds, and mixed sun exposure. Converting lawn to food production is an ideal no-till challenge. Begin with smothering techniques—cardboard, compost, and mulch—to suppress turf. Then layer in compost-rich soil, cover crops, and dense polycultures. Suburban growers benefit from better water access and fewer zoning constraints than urban growers, while still navigating compact spaces. Beds can be laid out along fence lines, between trees, or even on slopes with terraced mulched platforms.

In rural areas, land is not the limiting factor—labor and time are. Larger plots allow for deeper rotations, longer cover cropping periods, and full integration of grazing or multi-acre systems. Here, no-till might include large-scale cover crop mixes, roller crimpers, no-till seeders, and compost application by tractor. The principles remain: disturb less, feed biology, keep soil covered. Rural growers have the added challenge of scale—decisions affect more soil, more biomass, and more budget. However, rural no-till systems also yield massive soil-building potential when approached with biological care rather than mechanical force.

Across all settings, no-till adapts. From a two-gallon pot to a two-acre plot, soil care depends less on size and more on observation, patience, and biology.

Contaminated Soil Strategies

Safety is more important to soil health than nutrients alone. Lead, heavy metals, petroleum products, and other unidentified pollutants are frequently left behind in urban and post-industrial areas. For decades, suburban yards may have been treated with chemical fertilizers, pesticides, or herbicides. Chemical residues, industrial runoff, and spray drift can persist for years, so even rural regions are vulnerable. Understanding, not avoiding, is the first step in dealing with contamination for no-till growers.

Soil testing is the first step. Reliable labs can test for lead, arsenic, cadmium, and other heavy metals, along with pH and organic contaminants. Test both topsoil and subsoil where possible. Results will help you determine risk levels and mitigation needs.

Barrier gardening is a viable strategy in high-risk areas. Raised beds built with physical separation—geotextile fabric, concrete, or heavy-duty landscape cloth—allow you to grow in clean imported soil over questionable ground. Beds should be deep enough to support full root development, at least 12 to 18 inches, with clean compost and organic matter to build biology.

Phytoremediation uses specific plants to absorb or immobilize contaminants. Sunflowers, mustard, hemp, and certain grasses can uptake heavy metals from soil over time. These plants should not be composted or consumed—they are sacrificial. After several cycles, soil concentrations may lower enough to safely support edible crops.

Fungal remediation introduces fungi capable of breaking down hydrocarbons and binding metals. Inoculated wood chips, mycorrhizal inoculants, and mushroom compost may support early stages of cleanup in organically compromised sites.

Mulching and layering remain powerful tools. Thick mulch and compost layers reduce exposure to contaminated dust, reduce erosion, and provide a buffer between plants and polluted subsoil. Over time, biological activity can stabilize and compartmentalize toxins, especially in heavy-clay soils.

No-till does not mean ignoring contamination—it means working with care, building soil from the top down, and using biological allies to make unsafe soil safer. Every strategy begins with protection and ends with regeneration.

How to Scale Up or Down Without Losing Soil Health

No-till systems need to be adaptable. It is possible that what works on one raised bed will not work on five, and vice versa. Scaling the principles—not the processes—is crucial. Maintaining soil health necessitates consistency in strategy rather than equipment, whether extending into new terrain or reducing to a more manageable footprint.

Planning for systems, not simply beds, is the first step in scaling up. How will the compost be used? What is the plan for seeding and terminating cover crops? How will irrigation, rotations, and pathways be controlled? At a certain scale, hand instruments lose their efficiency, but their logic endures. Use a subsoiler instead of a broadfork. Use a compost spreader instead of a hand spreader. Instead of using scythes, use roller crimpers. Continue using mulching techniques, rotation schedules, and permanent bed layouts. When it imitates small-scale care with larger equipment, large-scale no-till flourishes.

At scale, cover crops become essential. In order to maintain soil cover, weed suppression, and fertility cycling, cash crops should be seeded with a multispecies mix. Broadcast spreaders, flail mowers, and no-till drills all aid in the minimal disturbance of huge land. Scaling up entails adopting the no-till philosophy to every field decision, never reverting to burning or plowing.

Reduction is also deliberate. Reduce increasing area if life requires it, but maintain your essential systems. Pay attention to a few beds with great efficiency. Continue applying mulch, cover crops, and compost. Relay cropping and interplanting might help you get the most out of small areas. For biological boosts without having to move bulky materials, use compost teas and worm castings.

Downsizing is a return to closeness, observation, and improved practice, not a sign of failure. Many of the most successful soil-building initiatives originate from modest, closely monitored plots. The secret is to appropriately scale your system for the energy and resources at hand, not to give up on scaling.

By adhering to biology, no-till adjusts to scale. Do not let it get exposed. Lessen the disturbance. Give it vitality. The same fundamental guidelines apply whether you are looking after a pasture, backyard, or rooftop. How imaginatively you use them makes a difference.

No-till farming is based on life-serving principles rather than size, location, or environment. The same qualities are desired by soil in both expansive farms and concrete courtyards: organic matter, protection, moisture retention, and unhindered movement. Asphalt jungles become areas of abundance thanks to urban farmers. Lawns are transformed into living systems by suburban growers. Farmers rehabilitate acres of degraded land into lush, productive fields.

No-till provides a route to ecological, economic, and biological regeneration in any situation. With tactics catered to possibilities and constraints, it meets growers where they are. It encourages innovation, flexibility, and long-term planning.

Contaminated locations require planning and caution, not hopelessness. With time, biology, and attention, even poisoned soil may recover. Infrastructure that aligns with your beliefs is necessary for scaling up. Clarity, accuracy, and humility are necessary while scaling down. Because no-till systems are based on relationships rather than control, they survive over time and space.

When growers stop asking, "How big is my plot?" they are successful. I begin to inquire, "What is the life of my soil?" Both can be answered by no-till. The ground beneath you, whether in a field, a backyard, or a pot, is waiting for your cooperation rather than more machinery. It will return more than you could have ever dreamed if you respect it, cover it, and build it. No-till farming promises that everywhere you grow.

Chapter 12:
From Passion to Profit—
The Business of Regenerative Growing

Startup Costs & Budgeting Tools

Although regenerative farming has its roots in soil, economics must also be considered. People enter the field with passion, but they stay there because of their planning. Establishing a no-till growing business needs more than just a desire to live off the land; it also calls for a thorough understanding of infrastructure, tools, capital, and the seasonal rhythm of costs and profits.

Fortunately, no-till systems frequently need less funding than traditional ones. There are fewer synthetic inputs to buy, no tractors to finance, and no plows or rototillers to maintain. Costs still accumulate, though, and they need to be considered up front. Immediate needs include infrastructure such as wheelbarrows, hand tools, tarps, seeds, compost, irrigation systems, and tarps. Materials for permanent beds, compost and additives, premium hand tools, seed trays, and pest control are usually included in a basic starter budget for even the smallest market garden. People switching from backyard gardening to small-scale farming can also need walk-in refrigerators, storage facilities, or delivery equipment.

Using budgeting tools helps keep costs under control throughout the launch's enthusiasm. Clarity is provided by a comprehensive spreadsheet that tracks both variable and fixed costs, such as seeds, compost, harvest containers, marketing, and insurance and utilities. Crop planning software, cash flow calendars, and Gantt charts are examples of tools that help you discipline your excitement. Do not forget to include contingency money for unforeseen circumstances. Without a financial cushion, a failing crop, a broken irrigation line, or a delayed compost delivery can completely disrupt operations.

Money is time, but security is also money. You can approach soil-building as an investment with both financial and ecological benefits if you are upfront about your initial commitment and what you will need to generate a profit.

Labor, Efficiency, and Smart Time Use

One of the most recurring expenses in any regenerative operation is labor. In order to adhere to the no-till philosophy—less waste, more intention, and maximum outcomes from least disturbance—labor must be optimized, whether you are working alone, with a partner, or with a seasonal workforce.

Efficiency in no-till systems is won by methods, not by taking short cuts. Tool placement, walkway access, and bed design must expedite tasks like mulching, transplanting, and compost distribution. Every season, permanent beds assist cut down on needless setup time. You can move swiftly and predictably thanks to modular tools, standardized bed lengths, and regular row spacing.

Every week, begin with a plan. Give top priority to the activities that improve soil and sustain yields, such as applying compost, moving plants, harvesting, and handling crops after harvest. Put off or assign the less important ones. It may seem efficient to plant all of your lettuce on one day, but it will take longer to harvest them all at once if you use crop plans to spread out your work over time.

The most labor-intensive aspect of the process may be harvest time. Invest in labor processes that minimize bending, lifting, and walking, as well as ergonomic equipment and wash station designs. Harvest zones should be situated adjacent to packing and storage areas. Bins and labeling tools with color coding. Efficiency is about lowering fatigue and friction, not about speed.

Getting assistance is a significant step. Start with jobs that are unique to a certain duty, like a compost delivery driver, a Saturday market assistance, or a morning harvest assistant. Consider labor to be an ongoing partnership. Give extensive training, recognize progress, and always relate the work to the goals of food justice and land maintenance.

The amount of work you accomplish is not the only factor in time utilization. It all comes down to executing the correct tasks with the appropriate equipment, in the right order, and at the right time. A tiny no-till grower can compete with much larger farms because of this.

Making No-Till Financially Sustainable

Regenerative growth has the potential to be a successful business strategy, not just a technique. However, achieving success necessitates balancing market realities with ecological integrity. Organic techniques are only one aspect of sustainability. It entails making enough money to continue without endangering your health or the land.

When no-till farmers learn to prioritize quality over quantity, they become financially successful. They prioritize nutrient richness, consistent flavor, long shelf life, and aesthetic appeal over producing the most yields. Premium pricing are demanded for high-quality crops, and consumers are willing to pay if the freshness, nutrition, and storyline justify the cost.

For small regenerative growers, direct sales are frequently the best option. Better margins than wholesale can be found through on-farm sales, restaurant partnerships, farmers markets, and community-supported agriculture (CSA) initiatives. Additionally, these markets foster client loyalty and partnerships. Every carrot, head of lettuce, and bunch of kale gains value when you inform your customers about soil health, flavor, and sustainability.

On a small scale With careful planning, no-till farms can flourish on as little as half an acre. Stacking functions—relay planting, interplanting, several successions, and maintaining full beds throughout the season—are key to success. Revenue per square foot increases with the number of turns per bed. The per-bed value increases even more when you include microgreens, herbs, or value-added goods (such as pickles, ferments, or flower bouquets).

Maintain a low overhead. Pay attention to labor-saving tools rather than only aesthetically pleasing ones. Give revenue-generating activities first priority. Keep an eye on the economics of your crops; some could be gorgeous, but at most they will break even. Others, such as herbs or salad greens, might provide significant benefits with little work and space.

Knowing what you need, making prudent financial decisions, and allowing the soil to eventually return your care are the keys to financial sustainability rather than accumulating more.

Try This On Your Plot: "Your Regenerative Startup Checklist"

Starting a regenerative growing business is equal parts clarity, courage, and compost. Here's a simple checklist to help you organize your thoughts, materials, and plans before diving in.

1. Assess Your Land and Water Access

☑ How much sun, slope, wind, and access does your space offer?
☑ What's the quality of the soil or imported growing medium?
☑ Is water available, pressurized, and reliable year-round?

2. Establish Your Growing Infrastructure

☑ Do you have permanent bed dimensions laid out?
☑ What are your pathways, fencing, and drainage solutions?
☑ Do you have compost, mulch, and amendments in place for startup?
☑ Are you using silage tarps or cardboard for initial bed prep?

3. Budget Your Core Tools and Materials

☑ Wheelbarrow, broadfork, seeder, harvest knives, and wash bins
☑ Irrigation system: drip or overhead (choose based on context)
☑ Shade cloth, insect netting, row covers for climate resilience
☑ Cold storage or harvest preservation tools

4. Plan Your Crop Calendar and Sales Channels

☑ Do you have at least 5–7 crops planned that can succeed in your climate?
☑ Are you growing for CSA, market, restaurants, or home use?
☑ What is your expected planting, harvest, and income timeline?

5. Set a Weekly Work Rhythm

- ☑ Which days will be for seeding, transplanting, and harvesting?
- ☑ How will you rotate tasks to avoid burnout and bottlenecks?
- ☑ Can you commit to a weekly review of labor, harvest, and soil care?

6. Define Your Soil Regeneration Plan

- ☑ How will you maintain cover crops, mulch, and compost inputs?
- ☑ Do you have a system for compost tea or microbial support?
- ☑ What indicators will you use to measure soil improvement?

7. Connect to Community and Mentors

- ☑ Join local growers' groups or cooperative networks
- ☑ Read regenerative farming books and listen to farmer-led podcasts
- ☑ Visit other no-till farms to learn systems that match your scale

Although this list is not all-inclusive, it serves as a guide to help you concentrate your efforts on the most important things: developing a business that respects both land and livelihood and growing well and sensibly.

It is not easy to go from passion to profit. It is a convoluted, dynamic journey influenced by soil, weather, market dynamics, and your personal drive. However, it is also a promising path. You can pursue regenerative growing without sacrificing your morals. It pushes you to hone them—to include your values into each instrument, transaction, and agricultural strategy.

Although startup costs are real, they may be controlled with imagination and clarity. Consistency, organization, and processes can all help to maximize labor. Intention leads to profitability: savvy marketing, tight crop rotations, and high-quality products. Most importantly, you must view your soil as an ally and business partner, not just a medium, if you want to succeed as a no-till farmer.

Healing the land is only one aspect of regenerative agriculture. It also aims to restore the connection between food and community, as well as between grower and labor. When the numbers add up, that healing becomes feasible. When you are supported by the systems rather than the other way around, it becomes sustainable.

Begin modestly. Increase gradually. Because the two are synonymous, invest in your soil just as you would in your future. And keep in mind that mission and profit are not mutually exclusive. It is what gives purpose longevity.

Chapter 14:
Soil-Smart Crop Planning

Choose Crops That Build the Soil

The demands of your market or your favorite vegetables are not the first steps in good crop planning. Your soil is the first step. Each crop has a unique manner of interacting with the soil, whether it pushes, pulls, feeds, or drains. By choosing crops based on the needs of the land rather than merely what the customer wants, you can adopt a soil-smart approach. You build a system that becomes more robust, more profitable, and more fertile over time when your crop plan nourishes your soil and your consumers.

Soil is actively built by certain crops. They shade the soil, produce biomass, inhibit weed growth, and promote biological activity. Some are naturally "extractive"—high feeders that need more water, nutrients, and care. These crops are perfectly OK, but they need to be balanced with regenerative companions.

Legumes, like beans, peas, and lentils, fix nitrogen through root nodules. They partner with rhizobia bacteria to pull nitrogen from the air and share it with the soil. These crops are soil builders, especially when their roots and residues are left in place.

Deep-rooted crops, such as daikon radish, sunflowers, or even tomatoes, can penetrate compacted layers and help aerate the soil. Their taproots open channels that later crops can follow, and they increase water infiltration.

Biomass producers like corn, sorghum, and squash offer generous above-ground material. When chopped and dropped or crimped as residue, they return carbon to the soil and foster fungal communities.

Low-impact greens—lettuce, arugula, spinach—may not build soil directly, but they work well in rotations, especially when planted into compost-mulched beds. They are quick-growing, cover bare ground, and do minimal damage if harvested properly.

Your crop list isn't just about food—it's about function. Choose crops that balance economic return with biological value. Grow the soil while you grow your income.

Balancing Yield and Soil Regeneration

The conflict between conserving the land and maximizing produce is something that every grower must deal with. Systems with high yields are alluring. They provide short-term revenue, draw in market clients, and fill CSA boxes. However, that high yield can quickly result in high depletion if rotation, biological input, and proper design are not implemented.

An incorrect decision is not imposed by soil-smart planning. It creates a cadence that makes regeneration and productivity compatible. Rotation is key, not only for crops but also for soil requirements.

Consider the following crop categories:

- **Heavy feeders:** Tomatoes, corn, cabbage, broccoli, and potatoes need rich, compost-fed beds. Follow them with soil builders.
- **Moderate feeders:** Peppers, carrots, onions, and beets require fertility but don't deplete the soil dramatically. Pair them with organic amendments and mulch.
- **Soil builders:** Legumes, oats, buckwheat, vetch, and clover enrich the soil, add biomass, and support microbial life. These crops don't make you money immediately, but they set the stage for better returns in the next cycle.

Instead of focusing only on yield, a regenerative grower balances it throughout the growing season. A cover crop or a short-season builder, such as bush beans, should come after a demanding crop, such as cauliflower or eggplant. Following a round of greens, top up the bed with compost and allow a rest crop to develop for six weeks.

A kale crop interplanted with lettuce or clover between early carrot rows is another option to think about. These combinations optimize space, shield the soil, and disperse the impact.

Balance does not just happen. It is an active component of the strategy. By keeping biology, diversity, and timing in mind, you aid the soil's recovery rather than "letting" it happen.

CSA, Market Garden, and Home Use Models

Crop plans vary depending on the growing model, but soil-smart thinking is beneficial for all of them. Your soil is shaped by how you design your beds, whether you are feeding a dozen households, supplying a weekly market, or simply filling your kitchen.

CSA (Community Supported Agriculture) models require diversity, consistency, and timing. Members expect a reliable box every week, filled with seasonal variety. This encourages the planter to use interplanting, staggered harvests, and succession planting to keep full beds throughout the year. Pressure is the soil problem here; beds are pushed aggressively and do not get much rest in between cycles. Compost mulch is used to buffer the load, fallow weeks are planned, and rapid soil builders like buckwheat are rotated in by astute CSA gardeners. Crop selection is balanced across beds and time and includes long-game crops (tomatoes, potatoes, and cabbage), mid-season mainstays (carrots, onions, and squash), and fast crops (radishes, salad mix).

Market gardens focus on income per square foot. Growers often specialize in high-value crops like arugula, kale, herbs, and tomatoes. The risk is overplanting heavy feeders and skipping regenerative support. In this system, every bed should have a fertility plan: compost topdressing, microbial inoculation, or a short fall cover crop. Rotate income crops with low-input crops like bush beans or mustards to reduce input costs and restore balance. Intercropping and dense spacing help maintain soil cover and microbial life.

Home use growers have the most flexibility—and often the greatest opportunity for soil health. You're not bound to deadlines or yield expectations. You can grow crops that feed the soil first, then the table. Experiment with perennial vegetables (like asparagus or rhubarb), integrate herbs and flowers for biodiversity, and rotate soil-friendly crops freely. Even a small home garden can become a model of regenerative abundance when crop planning centers on balance and biology.

No matter the model, crop planning is more than logistics—it's an invitation to collaborate with your soil. It's how you turn intention into a sustainable harvest.

A soil-smart crop plan lives on paper before it thrives in the soil. Use this simplified calendar framework to build your own regenerative rotation across the season.

Spring (March–May)

- Early soil-building options: oats, field peas, fava beans
- Quick cash crops: radish, salad mix, arugula
- Balanced crops: carrots, beets, turnips
- Cover crop to sow mid-spring for summer rest: buckwheat

Summer (June–August)

- Heavy feeders: tomatoes, peppers, squash, corn
- Soil maintainers: green beans, cucumbers, basil
- Mid-season soil builders: cowpeas, sunflowers
- Rotation prep: harvest early crops and seed fall cover crops by mid-August

Fall (September–November)

- Cool-season income crops: kale, spinach, lettuce, radish
- Soil builders for winter: rye, vetch, crimson clover
- Compost topdressing after last harvests
- Mulch bare soil or plant green manure mixes

Winter (December–February)

- Living roots where climate allows: garlic, overwintering onions, rye
- Passive recovery: mulch-only beds, fungal decomposition zones
- Planning season: review last year's nutrient flow, pest pressure, and bed use

Your soil will remain covered, varied, and constantly interacting with organic matter and living roots thanks to this calendar. Sort your crops by color according to their function (maintener, builder, or feeder), and try to switch up the functions in each bed throughout the year.

Make use of your calendar to help you make decisions. Make changes if you observe two heavy feeders trailing one another. Add a biomass crop or rest period if a bed does not already have one. The more planning you do now, the fewer inputs you will require later. Planning is preventive.

A key component of regenerative farming is soil-smart crop planning. It involves creating a dance between biology, harvest, and wellness rather than just picking veggies. Your farming is not a passive process for the earth. It takes part actively. A cycle of fertility, resilience, and abundance is created when your crop strategy nourishes both the land and the consumer.

Make deliberate crop choices. Let others reward, let others rest, and let some build. Strike a balance between restoration and yield. Plan your systems for growth across several seasons rather than just one season's profit. Your calendar, your compost, and your diversity are all instruments for a bigger purpose: a soil that improves rather than deteriorates with each passing year.

Let your crop plan reflect your ideals, whether you farm for your own plate, for markets, or for families. Allow it to replenish, restore, and regenerate. Allow your beds to convey not only what you planted but also your level of care. It is therefore soil-smart. It is therefore regenerating.

Chapter 9:
Pest and Disease Resistance
Starts in the Soil

Healthy Soil = Resilient Plants

The soil is the cornerstone of any successful garden. This unseen layer of soil beneath our feet is a living, breathing ecosystem that directly affects the health, vitality, and disease and insect resistance of plants. It is more than just a place to anchor roots. Microbial life, fungi, bacteria, nematodes, and earthworms are abundant in healthy soil, and they all work together symbiotically to break down organic matter and release vital nutrients. By creating a biological barrier, this tiny workforce drives out dangerous diseases before they ever get to your plants.

Rich in organic matter, soil retains more water, has a better texture, and drains more effectively. Loamy soil, which is rich in humus and balanced in sand, silt, and clay, is frequently regarded as the ideal growing medium. This structure prevents waterlogging, which frequently results in bacterial and fungal infestations, while allowing roots to breathe and spread out. This rich matrix is created by compost, aged manure, leaf mold, and worm castings, which guarantees that your garden soil is not only alive but flourishing.

The unsung hero of disease and pest resistance is microorganisms. For instance, mycorrhizal fungi encircle plant roots in a protective web that enhances nutrient uptake and acts as a barrier against soil-borne diseases such as Pythium and Fusarium. As a natural defense mechanism, other microorganisms such as Bacillus subtilis and Trichoderma species actively reduce fungal spores and compete with pathogenic bacteria. There is a far lower chance of disease outbreaks and insect incursions when this soil food web is in equilibrium.

Furthermore, nutrient-dense plants thrive in healthy soil. Stronger cell walls are produced by plants with balanced mineral profiles, particularly sufficient calcium, potassium, and magnesium. Insects find it more difficult to chew through these tougher tissues, and fungi that are attempting to colonize leaf surfaces find them less appetizing. Additionally, a nutrient-rich plant produces fewer stress signals, such as ethylene, which are used as homing beacons by many pests. For many frequent hazards, a well-fed plant grown in biologically active soil essentially becomes invisible, or at least unappealing.

Natural Pest Deterrents and Companions

One of the best strategies to keep pests away is to use the strength of nature's own mechanisms. A tried-and-true technique called companion planting entails growing specific plants close to one another to promote growth or offer pest protection. Marigolds, for instance, release thiophenes into the soil, which are substances that discourage nematodes and other pests that live in roots. Whiteflies and hornworms are repelled by basil planted next to tomatoes. As sacrifice plants, nasturtiums draw pests away from susceptible crops like broccoli and kale.

Strong odors from aromatic herbs like lavender, mint, and rosemary can fool insects. These herbs are especially effective at garden bed borders, creating a sensory barrier that confuses pests including flea beetles, cabbage moths, and aphids. Additionally, by upending monoculture design, interplanting makes it more difficult for pests to find and infest host crops. Pests that depend on smell and sight to locate their favorite food sources become confused by the variety in the garden.

Another good tactic is to support natural predators. Ground beetles, parasitic wasps, ladybugs, and lacewings all eat pests. Planting a range of nectar- and pollen-producing flowering plants, such fennel, dill, and yarrow, will help create a friendly habitat for these allies. Beneficial insects can find safe places to mate and hibernate in the garden, even if weed patches are allowed to flourish in less-traveled areas.

In addition to insects, frogs and birds are effective pest-control friends. Another line of defense is to welcome these species into your yard by installing birdhouses or water features. Allowing hens to occasionally wander along garden walkways can help control insect and larval numbers without endangering established plants.

Non-Toxic Integrated Pest Management

Prioritizing prevention and natural controls over chemical interventions, Integrated Pest Management (IPM) is a methodical, multi-layered approach. Understanding the lifecycles and behaviors of pests is the first step towards non-toxic IPM. You can schedule your interventions to be most successful and least disruptive by keeping an eye out for patterns, such as when aphids multiply or cabbage moths lay eggs.

Physical control is the initial stage in non-toxic IPM. Row covers let light and moisture in while shielding seedlings from insects. Picking pests by hand, such squash bugs or tomato hornworms, can be time-consuming but efficient. Yellow sticky traps provide you with early warning of population booms by capturing flying insects such as fungus gnats and whiteflies.

Cultural customs are also very important. Crop rotation stops disease accumulation in the soil and interferes with pest breeding cycles. Adequate plant spacing promotes healthy air circulation and lowers the risk of fungal outbreaks. Early morning watering prevents mildew and decay by allowing the leaves to dry before dusk. In addition to keeping the garden neat, pruning sick leaves and getting rid of infected plant detritus prevents diseases from spreading.

The foundation of non-toxic IPM is biological control. Ecological equilibrium is restored by introducing beneficial insects, either by buying them or by luring them in spontaneously. Root weevils and grubs can be targeted by adding nematodes to the soil. A naturally occurring bacteria called Bacillus thuringiensis (Bt) is safe for mammals and birds but efficient against caterpillars. The neem tree yields neem oil, which interferes with insect hormones to prevent growth and reproduction. Diatomaceous earth, which is derived from fossilized algae, dehydrates insects and scrapes their exoskeletons; however, it should only be used sparingly because it can also damage beneficial insects.

In IPM, monitoring is essential. Keep note of weather, plant health, and insect observations with a journal or garden app. Patterns show up over time, and interventions becoming more accurate. Gardeners can target particular pests during their most vulnerable stages instead of spraying widely at the first hint of difficulty. This accuracy reduces unintended harm to the environment and beneficial creatures.

Myth vs. Reality: "Organic Sprays Are Always Safe"

A widespread misunderstanding among gardeners is that a product must always be safe just because it is branded as "organic." Despite coming from natural sources, organic sprays do carry some hazards. For example, chrysanthemum blooms are used to make pyrethrins, which are plant-based pesticides. They are broad-spectrum and natural, but they also kill beneficial insects like ladybugs and bees as well as pests. Pollinator numbers can be severely reduced by the misuse of pyrethrins, particularly during bloom season.

Even neem oil, which is highly regarded for its adaptability, has limitations. Neem can burn leaves if it is sprayed during the strongest sunlight. A covering that prevents photosynthesis may be produced by heavy, repeated treatments. Additionally, neem's hormonal interference can interfere with beneficial insects' reproduction cycles in addition to harming pests.

Another common organic alternative is copper-based fungicides, which gradually build up in the soil. Excessive copper undermines the long-term health of soil by harming earthworms and microorganisms. Similarly, if inhaled during application, sulfur sprays can irritate both human and animal respiratory systems. When abused, they can also change the pH of the soil, which can impact the availability of nutrients for plants.

It is deceptive to assume that "natural equals harmless." Context, dosage, timing, and the general health of your garden ecosystem all affect safety. After biological and preventative controls, any intervention, whether synthetic or organic, should be used as a last option. Wearing protective gear, reading instructions carefully, and avoiding spraying in windy or extremely hot conditions are all crucial. More importantly, pay attention to how your garden reacts and make the necessary adjustments. When using organic treatments responsibly, one must strike a balance between short-term need and long-term soil and ecosystem health.

The overuse of organic products can result in resistance, just like with synthetic chemicals. Gardeners may be forced into a loop of ever-increasing interventions if pests are continually exposed to the same control strategy because they may acquire immunity. To prevent any one product from becoming the mainstay of your pest management strategy, rotate your remedies, use them sparingly, and incorporate additional techniques.

Although organic solutions can be effective instruments, they need to be used with the same caution, diligence, and knowledge as any other chemical. The secret to making sure they benefit your garden rather than hurt it is to be careful about how, when, and why you use them.

Resistance to pests and diseases does not begin with devices or sprays. It starts at the intersection of biology, chemistry, and good stewardship, deep within the soil. Creating a thriving soil environment creates the foundation for robust, healthy plants. Gardeners can create a thriving environment in addition to thriving plants by using natural insect deterrents, promoting beneficial companions, using non-toxic IPM techniques, and remaining skeptical of so-called "safe" treatments. A garden is a dynamic community where every decision either boosts or depletes the immune system. It is more than just a collection of plants. We select the route of sustainability, abundance, and ecological harmony when we decide to start with the soil.

PART 5:
The Grower's Life —
Long-Term Stewardship

Chapter 16:
Your No-Till Year

Seasonal Soil Tasks

No-till gardening is a way of life that aligns with the cycles of nature, not just a technique for managing soil. No-till techniques use organic inputs, observation, and light intervention to cultivate the land all year round rather than altering the soil structure with each season. Consistent and intentional seasonal soil maintenance is necessary for this strategy to succeed.

The emphasis switches to gradually awakening the earth in the spring. Wait until the earth has warmed up before removing winter mulch. This gradually exposes soil life to sunlight and warmth while assisting in retaining the moisture that has gathered throughout the colder months. Without disturbing the soil layers underneath, cover existing mulch with a new layer of compost or worm castings. This reactivates microbiological life and replenishes nutrients. Do not dig up weeds if they appear. To prevent regrowth, cut them off at the root and cover them with a thick layer of mulch.

Summertime calls for alertness. High temperatures and rapid decomposition can upset the equilibrium of soil life. To control soil temperature and stop evaporation, keep your soil covered with organic mulch like straw, chopped leaves, or shredded wood. Use fish emulsion or compost teas as a top dressing if fertility seems low. These liquid feedings will not disrupt the soil while yet supplying vital nutrients. Keep a close eye on pest and disease activities because they are frequently linked to water imbalances and soil health.

The emphasis shifts to replenishing in the fall. Lay a thick layer of completed compost and plant cover crops such as vetch, rye, or clover. Once trimmed back, these living mulches add organic matter, inhibit erosion, and control weeds. The decomposing plant creates a natural mulch layer for the upcoming months if winter-kill crops, such as oats, are used. Additionally, sick plant matter can be removed from the surface in the fall without removing the roots. Let roots break down naturally to provide food for soil organisms.

Although winter is a time of rest, significant work is still being done underneath the surface. Organic debris is still being broken down by soil bacteria. Worms mix the layers and tunnel. Under snow and mulch, fungal networks continue to function. This calm time is crucial. To avoid compaction, do not walk on garden beds during this period. Use straw or evergreen boughs to shield the soil's surface from wind and frost if there is not much snow.

Instead of digging and disturbing, the objective is to layer and feed throughout each season. The structure, resilience, and microbial richness of the soil are all improved with each organic addition. No-till gardening respects the earth as a living thing that may flourish with time, patience, and careful attention.

Monthly Calendars

A no-till year becomes more manageable when broken into a month-by-month system. By following a natural rhythm, each task builds on the last.

January: Review last season's notes. Sketch out bed rotation plans and select crops for the year. Order seeds and cover crop varieties suited for your climate and goals.

February: Test soil if needed. Start collecting organic materials for compost. Plan greenhouse or indoor starts for early crops. Build or mend raised beds without disturbing the existing soil.

March: Sow cold-hardy seeds indoors or in protected outdoor areas. Begin warming beds with plastic sheeting or cloche covers. Avoid disturbing mulch too early—soil life still needs insulation.

April: Remove winter-killed cover crops by cutting them at the base. Add fresh compost directly on top of old mulch. Begin direct seeding for carrots, radishes, peas, and greens.

May: Transition transplants to beds. Monitor moisture levels closely—mulch should prevent drying but still allow rain to filter through. Add compost teas to stimulate soil biology. Spot-mulch any open areas.

June: Prune, harvest, and observe. Thin overgrown areas without digging. Reapply mulch where needed. Consider succession planting. Avoid leaving any soil surface bare.

July: Focus on maintenance. Watch for pests and diseases. Add seaweed or fish-based foliar sprays to boost plant health. Begin light composting again with garden waste and food scraps.

August: Harvest heavily and prepare fall crops. Start clearing spent summer plants by cutting at the base. Sow fast-growing cover crops in empty beds. Keep soil shaded and moist.

September: Plant garlic, onions, and overwintering crops. Add another compost layer across all beds. Cut back summer crops. Leave roots in the ground. Let vines and foliage compost in place.

October: Sow winter cover crops if not done earlier. Mulch deeply around perennials. Clean up only diseased material. Layer beds with chopped leaves or wood chips for winter protection.

November: Reflect and document. Note what worked and what didn't. Turn compost piles but avoid soil turning in beds. Continue mulching. Store tools, sharpen blades, and organize supplies.

December: Let the soil rest. Replenish bird feeders and install habitat for beneficial insects. Snow cover acts as natural insulation. Use this quiet month for learning, reading, and planning.

Each month in a no-till garden reinforces long-term soil health. By tracking tasks seasonally, the soil becomes more predictable, nutrient-dense, and biologically active with every passing year.

Preparing Beds for the Next Cycle

No-till beds do not require turning over, digging, or flipping at the conclusion of a growing cycle. They need to be layered, rested, and the soil's current life must be preserved.

Harvest the crops at the base first. Stay away from uprooting whenever you can. There are several reasons to leave roots unbroken. Roots make passageways for water and air as they decompose. They also provide food for bacteria and fungus that get their energy from plant sugars. This subsurface food network is disrupted when roots are removed.

Next, evaluate the layer of mulch. Add a new layer of straw, shredded leaves, or grass clippings if it has become very thin or decomposed. These materials ought to be pesticide-free and clean. Mulch provides the microbial life that maintains fertility in addition to shielding the soil from erosion and compaction.

Apply two to three inches of aged manure or compost on top of the entire bed. During the months when the soil is dormant, this serves as a covering to protect and nourish it. Allow gravity, earthworm activity, and natural rainfall to gently incorporate it into the soil profile instead of mixing it in.

When it comes to no-till preparation, cover crops are essential. While grains like winter rye contribute biomass and stop erosion, green manures like crimson clover or field peas fix nitrogen. To guarantee that the soil is never barren, plant cover crops in the fall or early spring, depending on your climate. These covers are grown, then clipped at the base and left as mulch. Their roots remain in place, breaking down and forming underground structures.

Choose sheet mulching if cover crops are not practical. Apply cardboard in layers, then compost, and finally organic mulch. In addition to feeding soil bacteria and suppressing weeds, this technique facilitates a seamless transition to the following planting cycle. It works particularly well in places where the soil is compacted or neglected.

Gentle preparation is also beneficial for perennial beds. Remove any dead growth and either compost it separately or leave it as mulch. To mitigate temperature fluctuations and avoid frost heave, reapply mulch. Perennials should not have their crowns or root zones disturbed so they can continue to form deeper networks every year.

In a no-till system, preparation is a silent act of support. It is more important to nourish the soil, cover it, and have faith in the biological processes that are currently in place than it is to rearrange the soil. With each cycle, your garden gets more resilient and productive the less you disturb it. Each season builds on the previous one, becoming stronger, richer, and more in sync with the natural rhythms, rather than beginning anew.

Chapter 17:
Voices from the Field

Case Studies from Urban and Rural Growers

Although the lessons of the soil are universal, different locations will experience them differently. Growers experiment, adapt, and change as they go from little balcony gardens to large homesteads. These case studies highlight a variety of actual soil stewards, including both rural traditionalists and urban innovators, who prove that soil health is determined by intention and care rather than acreage.

Urban Grower: Jasmine Reyes, Detroit, MI

Jasmine oversees a 600-square-foot backyard garden in a community with a high population density. She began by creating raised beds on dirt that had been compacted and covered in concrete scraps. She used lasagna layering—cardboard, compost, straw, and shredded leaves—to implement no-till techniques over a three-year period. Finding high-quality organic materials in the city was her biggest obstacle. Jasmine transformed waste into wealth by establishing neighborhood composting initiatives and collaborating with nearby coffee businesses to provide grounds and brewers to provide leftover grains. Her yields now equal those of conventional plots, and her garden is used as a learning environment for young people in her community.

Rural Grower: Harold McLeod, Asheville, NC

After decades of soil degradation, Harold took over his 10-acre Appalachian farm and switched from mechanical tilling to regenerative no-till techniques. He describes fighting against nutrient leaching, compacted clay, and erosion. He has gradually rebuilt organic matter and water-holding capacity by returning native perennials, incorporating silvopasture rotations, and planting deeply rooted cover crops. In addition to rotating grazing for sheep and poultry, Harold now plants heirloom vegetables, demonstrating how sustainable principles can support a variety of activities.

Real Soil Wins and Losses

No growing journey is without setbacks. Soil, like any living system, teaches through success and failure. Growers who embrace observation and adaptability often reap the long-term rewards.

Win: From Dead Dirt to Living Earth

Karen Ng, a rooftop grower in Vancouver, started with sterile bagged soil. Her first year yielded pale lettuce and stunted carrots. After learning about compost tea, mycorrhizal inoculation, and mulch layering, she rebuilt her substrate from the top down. Within two seasons, microbial counts surged, earthworms began appearing in planters, and her vegetable harvests doubled. She now inoculates each new container with compost and refuses to disturb her layers, noting that "once the life is there, you protect it like treasure."

Loss: Too Much of a Good Thing

Miguel Torres, a one-acre farmer in New Mexico, delivered a crucial warning. Without taking into account the rates of decomposition, he eagerly added layers of compost and raw manure. His nitrogen-rich soil produced poor fruit development but rapid vegetative growth. Even worse, some incomplete compost exposed vulnerable crops to root rot and fungus gnats. Miguel made the necessary adjustments by using well-aged compost, distributing brown and green components evenly, and adding organic matter gradually. Although his crops recovered, he stresses the importance of honoring the timing and maturity of the soil: "Feed your soil—but let it chew slowly."

Testimonial Spotlight: First-Person Grower Quotes

Real voices carry lessons data cannot. These direct quotes from growers highlight the emotional and philosophical shifts that occur when one commits to working with soil instead of against it.

"I used to think I was feeding plants. Now I know I'm feeding an entire underground city. The plants are just the part we see."
— Jasmine Reyes, Detroit, MI

"Switching to no-till wasn't just about saving labor. It was about breaking a cycle. I realized I was erasing the memory of the soil every time I turned it over."
— Harold McLeod, Asheville, NC

"The first time I saw my soil crumble like cake, I cried. I never thought rooftop containers could hold so much life. It was a miracle made from compost."
— Karen Ng, Vancouver, BC

"My biggest failure taught me restraint. Soil doesn't need a buffet—it needs a slow-cooked meal."
— Miguel Torres, Albuquerque, NM

"The best thing I ever did for my garden was stop rushing it. I leave roots in the ground, mulch the beds, and let the worms work overnight. I sleep better knowing I'm not interrupting the process."
— Lena Davis, Portland, OR

"Before, I was chasing perfection with chemicals. Now I aim for balance. A few bugs, a few weeds, but the soil—rich, dark, living—is the real victory."
— Zahid Rahman, Leicester, UK

Each testimony underscores that soil stewardship is as much about mindset as it is about method. These growers speak not as technicians, but as guardians of life beneath the surface. Their words carry humility, persistence, and joy—traits earned through hands-on, long-term work with the land.

"Voices from the Field" reminds us that while guides and techniques are valuable, the most powerful education comes from those who have walked the path. These stories, from rooftops to rural ridges, prove that good soil isn't about location—it's about connection. These growers didn't just change their soil—they let the soil change them.

Each of them teaches that resilience is not born from force but from listening. By observing nature, respecting timing, and staying consistent, they transformed barren dirt into thriving ecosystems. Their work honors the soil not as a tool, but as a partner in growth, a keeper of memory, and a teacher of patience.

Chapter 18:
The Future is Soil—
Activism, Climate, and Regeneration

Soil and Climate Action

Soil, a potent ally in the battle against climate change, is found beneath every flourishing crop, garden, or forest. This living system can buffer temperature extremes, control water cycles, trap and store atmospheric carbon, and sustain biodiversity. Regenerative soil management transforms soil into a growth medium and a first line of defense against the planet's worsening environmental disaster.

Carbon is stored in soil by photosynthesis. After absorbing carbon dioxide, plants transform it into sugars, which are then released through their roots to nourish soil microorganisms. The carbon in humus and organic materials is then stabilized by these microbes. This technique is long-lasting, effective, and natural. Thousands of pounds of carbon can be stored each year in one acre of good, biologically active soil, keeping it below and out of the atmosphere.

On the other hand, degraded soil releases carbon. Synthetic fertilizers, tillage, and chemical-intensive methods degrade organic matter, speed up erosion, and disturb microbial life. Stored carbon oxidizes and returns to the atmosphere when soil is cracked and exposed. This erodes the fundamental basis of food production in addition to increasing greenhouse gas concentrations.

One of the most economical climate remedies is to restore the health of the soil. Soil regeneration does not require complex infrastructure, factories, or fuel, in contrast to engineered carbon capture solutions. Compost, cover crops, varied planting, and human attention are all necessary. Every compost pile, mulched walkway, and no-till bed is a climate mitigation measure. Climate work is soil labor.

The Role of Growers in a Regenerative Future

Growers are in a unique position to further the regenerative movement since they are farmers, gardeners, and land stewards. Every day, they engage with the land. They make choices that have an impact on carbon balances, water systems, and local environment. Growers influence the future of soil with each season, each shovel, and each seed, whether they are tending to backyard beds or hundreds of acres.

The regenerative gardener concentrates on methods that enrich soil rather than diminish it. This covers rotational grazing, composting, polyculture plantings, perennial crops, and no-till techniques. They promote natural methods of nutrient cycling and pest control by eschewing synthetic fertilizers and pesticides. They support fungal networks that hold soil particles together and nourish new life by allowing roots to remain in situ.

Additionally, growers serve as activists and instructors. They have an impact on communities when they share what they have learnt via social media, local workshops, school gardens, or farmers markets. When one neighbor notices another's rich, colorful plot and inquires, "How did you grow that?" the transition from extraction to regeneration frequently starts. The seed for change is in that inquiry.

Growers play more than just a food function. Regenerative farming enhances watershed health, restores bird habitats, and boosts pollinator numbers. Rainfall is absorbed and filtered rather than pouring off and bringing pollutants into rivers when the soil is protected and there is a wealth of vegetation. Crop producers are not the only things that growers do. They act as both cultural change agents and ecosystem stewards.

What You Can Do Today, No Matter Your Scale

To participate in the soil solution, you do not have to own a farm. There are worthwhile activities accessible at every level, whether you are caring for a suburban lawn, a city balcony, or a rural pasture.

Feed the soil first. Instead of using commercial fertilizer, use compost. Instead of using bare dirt, use mulch. Let clover grow in between rows or plant cover crops. Allow spent plants to break down naturally. Reduce the amount of disturbance by not rotating, tilling, or disturbing the carefully constructed environment.

Even a tiny section of your lawn can be transformed into wildflowers, herbs, or natural grasses. These deeply rooted species help pollinators, use less water, and capture more carbon. Rather than bagging certain leaves, let them break down organically. When you let the land under your feet take part in its own cycle of renewal, it flourishes.

Gather rainfall and apply it to your garden. This helps restore groundwater supplies and eases the strain on nearby water systems. Prioritize water-wise plants and soil-building methods that enhance retention, like lasagna layering or hugelkultur, in regions that are prone to drought.

Speak up. Encourage regenerative agriculture on a national and local scale. Encourage laws that incentivize conservation, provide funding for studies on soil health, and shield land from excessive development. Participate in community-supported agriculture (CSA) initiatives that share your ideals. Purchase from growers who employ regenerative farming methods. Every purchase is a vote for a resilient, care-based, and biodiversity-based future.

Educate others as well as oneself. Ask questions, volunteer, read, and watch. Launch a composting campaign at your place of employment or education. At a community garden, provide a workshop about soil. Promote regeneration-focused programs at your neighborhood library or environmental organization.

Being involved in the regenerative movement is more important than being flawless. When handled with respect and purpose, even a tiny piece of land can aid in the healing of the entire planet.

Callout: "Join the Movement"

Soil is not the only issue here. It is about changing the way we view food, the land, and one another. The urge for regeneration is to shift from extraction to reciprocity. From immediate yield to long-term care. From individual development to group resilience.

Take a look at your earth outside and join the effort. Examine what is possible, what is missing, and what is alive. Start the process of creating, nourishing, and safeguarding it. Next, take another. Talk about what you discover. Pay attention to what other people are doing. Do not merely build beds; build networks. Alongside kale, cultivate knowledge. Grow justice with tomatoes.

Growers, diners, kids, seniors, activists, scientists, poets, and citizens are all welcome in our movement. In every way, soil is common ground. Our future—not only of food, but also of air, water, and climate—lies in its particles. Our hope is in its complexity. And we are accountable for its upkeep.

There is no trend in regeneration. This is a tipping point. A tribute. a reunion with the things that have kept life going.

PART 6:
Grower's Toolbox

A knowledgeable farmer uses information, instruments, and reference materials to help them make well-informed decisions in the garden rather than merely trusting their gut feelings. Part VI aims to provide you with just that: a practical, overview toolkit to support your no-till, regenerative path. To help you improve your system and expand your soil knowledge, these appendices include printable worksheets, soil insights, cheat sheets, and real-world resources.

Appendix A: Sample Soil Test Results & Interpretations

Knowing what is in your soil is the first step to understanding it. A soil test determines the cation exchange capacity (CEC), organic matter percentage, pH, and nutrient levels of your soil. This appendix's sample report provides explanations for each value and displays typical test findings from a transitional no-till garden.

- **Soil pH:** The ideal range for most vegetables is between 6.0 and 7.0. This test sample reads 6.4—great for nutrient uptake.

- **Organic Matter (OM):** This test reports 5.2%, a healthy number indicating active microbial life and good soil structure.

- **Nitrogen (N):** Often low in tests because it is mobile. The sample shows "Low," which means your composting and legume rotations should be emphasized.

- **Phosphorus (P):** "Sufficient," but excessive levels can pollute water systems. Maintenance doses only.

- **Potassium (K):** "Deficient," calling for inputs such as greensand, kelp meal, or composted banana peels.

- **Calcium (Ca) and Magnesium (Mg):** Present but slightly out of balance, which may affect soil texture and plant vigor.

Also included are visuals for interpreting bar graphs and color-coded values from popular soil labs, plus notes on how to choose between basic and comprehensive soil tests. These results guide soil amendment decisions, crop choices, and whether your soil ecosystem is on the right regenerative path.

Appendix B:
Soil Amendment Cheat Sheet

This section is a quick-reference chart organized by nutrient, amendment type, and application guidelines. Whether your soil needs more nitrogen or is showing signs of calcium deficiency, this cheat sheet will direct you to the best organic amendments.

Nitrogen (N) Sources:

- Feather meal – slow-release
- Blood meal – quick-release
- Alfalfa meal – moderate release + microbial stimulant
- Legume cover crops – seasonal boost

Phosphorus (P) Sources:

- Rock phosphate – long-term, slow release
- Bone meal – more available form
- Compost – moderate levels and good for microbial activity

Potassium (K) Sources:

- Kelp meal – gentle and adds trace minerals
- Greensand – slow release, good for sandy soils
- Wood ash – use sparingly to avoid pH spikes

Calcium (Ca) and Magnesium (Mg):

- Gypsum – adds calcium without raising pH
- Dolomitic lime – supplies both calcium and magnesium
- Epsom salts – fast-acting magnesium source

Trace Minerals:

- Azomite – a wide spectrum of trace elements
- Seaweed extract – micronutrients + plant hormones

Each entry includes average application rates per 100 square feet, ideal timing (pre-planting, mid-season, etc.), and whether the material can be applied as a tea, topdress, or soil drench. This cheat sheet keeps your soil support straightforward and manageable.

Appendix C:
Crop Rotation Planning Worksheets

Crop rotation is essential in a no-till system. It helps disrupt pest life cycles, reduce disease, and balance soil nutrients. This appendix includes printable and digital worksheets to map out your crop succession plans over multiple years.

Each worksheet includes:

- Bed Number or Name
- Current Year's Crop(s)
- Crop Family (e.g., Brassicaceae, Solanaceae)
- Soil Impact (e.g., heavy feeder, nitrogen fixer, light feeder)
- Following Year Plan
- Notes for Companion Plants or Cover Crops

A legend helps identify which families should not follow one another and which cover crops best suit each sequence. There are also examples of 3-year and 4-year rotation templates using common crops like tomatoes, kale, beans, and carrots.

Sample scenario:

- Year 1: Tomatoes (Solanaceae) – Heavy feeder
- Year 2: Peas (Fabaceae) – Nitrogen fixer
- Year 3: Carrots (Apiaceae) – Light feeder
- Year 4: Kale (Brassicaceae) – Heavy feeder + cover crop mid-season

Whether you manage four beds or forty, these worksheets help visualize your planting rhythms and maximize soil health through smart, deliberate planning.

Appendix D: Glossary of Key Terms

Clear understanding of terminology empowers action. This glossary defines common and complex terms used throughout the book in clear, no-jargon language. Whether you're a new grower or brushing up, this section ensures you speak the language of soil fluently.

Examples:

- **Aggregate** – Clusters of soil particles that form crumb-like structures essential for water retention and air movement.

- **Biochar** – Charcoal-like substance made from organic material; used to improve soil structure and microbial habitat.

- **Cation Exchange Capacity (CEC)** – A measure of how well soil holds onto nutrients; higher CEC means better nutrient availability.

- **Cover Crop** – Plants grown primarily to protect and improve soil, not for harvest. Examples include clover, rye, and buckwheat.

- **Fungal-Dominant Soil** – Soil with a higher fungal population than bacterial; often favored by perennials, trees, and shrubs.

- **Green Manure** – A cover crop grown and then cut down to decompose in place, enriching the soil with organic matter.

- **Inoculant** – A substance containing beneficial microorganisms used to boost soil biology or seedling resilience.

- **Tilth** – The physical condition of soil in terms of its suitability for planting. High tilth means loose, well-aerated soil.

- **Vermicompost** – Nutrient-rich compost produced using worms, particularly red wigglers, prized for plant health and soil life.

The glossary closes the loop on the Grower's Toolbox by building confidence in interpretation and vocabulary, creating a firm foundation for both conversation and cultivation.

www.ingramcontent.com/pod-product-compliance
Lightning Source LLC
Chambersburg PA
CBHW052338210326
41597CB00031B/5300